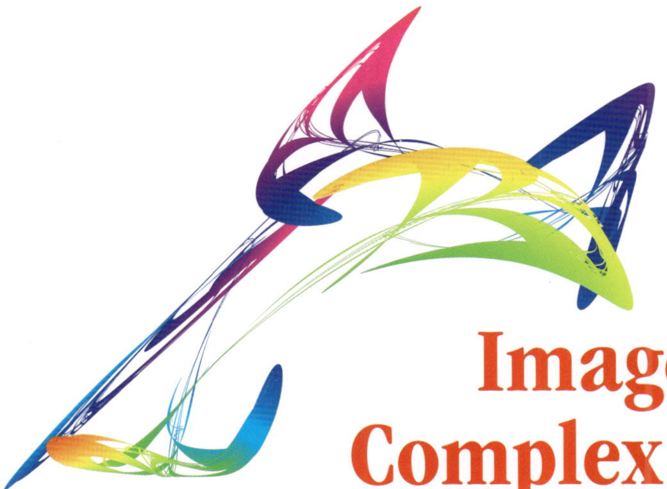

Images of a Complex World

The Art and Poetry of Chaos

Robin Chapman
Julien Clinton Sprott
University of Wisconsin-Madison, USA

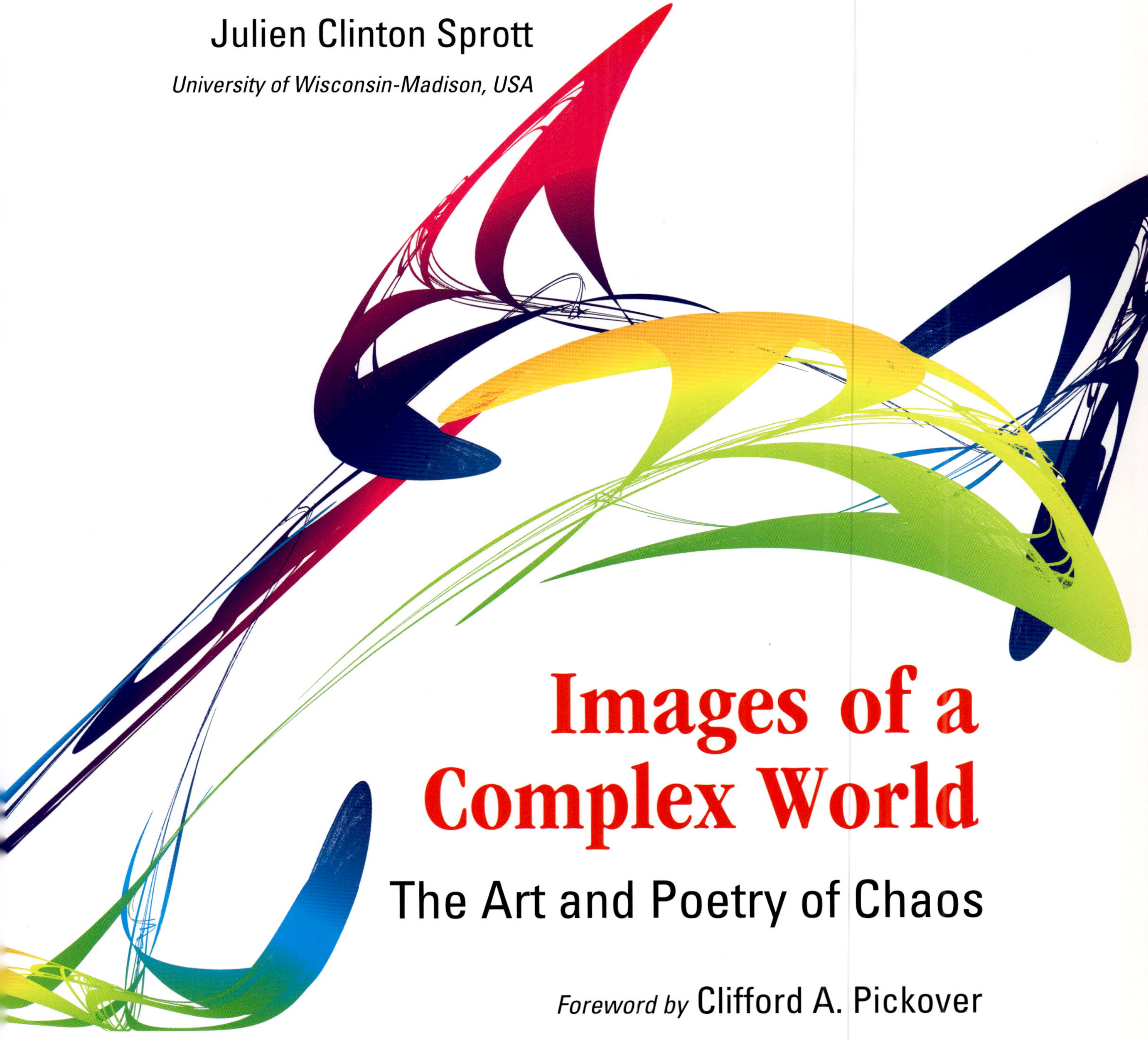

Images of a Complex World

The Art and Poetry of Chaos

Foreword by Clifford A. Pickover

World Scientific

NEW JERSEY • LONDON • SINGAPORE • BEIJING • SHANGHAI • HONG KONG • TAIPEI • CHENNAI

Published by

World Scientific Publishing Co. Pte. Ltd.

5 Toh Tuck Link, Singapore 596224

USA office: 27 Warren Street, Suite 401-402, Hackensack, NJ 07601

UK office: 57 Shelton Street, Covent Garden, London WC2H 9HE

Library of Congress Cataloging-in-Publication Data

Robin S. Chapman

Images of a complex world : the art and poetry of chaos / Robin Chapman, Julien Clinton Sprott ; foreword by Clifford A. Pickover.

p. cm.

Includes bibliographical references and index.

ISBN 981-256-400-4 -- ISBN 981-256-401-2 (pbk.)

1. Dynamics. 2. Chaotic behavior in systems. 3. Fractals. 4. Nonlinear theories. 5. Mathematics in art. 6. Mathematics in literature. 7. Digital art. I. Chapman, Robin S. II. Sprott, Julien C.

QA845.C46 2005

003'.857--dc22

2005042402

British Library Cataloguing-in-Publication Data

A catalogue record for this book is available from the British Library.

Printed by FuIsland Offset Printing (S) Pte Ltd, Singapore

Foreword

I wonder whether fractal images are not touching the very structure of our brains. Is there a clue in the infinitely regressing character of such images that illuminates our perception of art? Could it be that a fractal image is of such extraordinary richness, that it is bound to resonate with our neuronal circuits and stimulate the pleasure I infer we all feel. — Professor Peter W. Atkins, Lincoln College, Oxford University, "Art as Science," *The Daily Telegraph*

Chaos and fractals are revolutionary topics that find important applications in science, art, mathematics, and even philosophy. Professor Clint Sprott has been at the center of this cyclone since the early 1990s, and together with his coauthor Professor Robin Chapman, they have produced a unique visual and poetic survey of its manifold consequences. No one should miss the experience of stepping through the portals of this beautiful book into fantastic new worlds that computers are now exploring in the way a microscope explores the awesome wonders of nature. The eclectic coauthors combine logic and language as they present an array of art forms sure to stimulate your imagination and sense of wonder at the incredible vastness of our mathematical universe.

Doctors Sprott and Chapman are well suited to the task of conveying the wonders of fractals. Sprott, a Professor of Physics at the University of Wisconsin, has published hundreds of scientific papers on plasma physics, chaos, fractals, and complexity, and he has written several books on topics ranging from chaos to electronics to numerical recipes. He is also a well-known science popularizer with his *Wonders of Physics* program, where he demonstrates physics to a wide audience.

Dr. Chapman has a bachelor's degree in psychology from Swarthmore College and a Ph.D. from the University of California at Berkeley. Her current research focuses on language and cognitive development in children and adolescents. Chapman's poems have appeared in various journals and books, and her poems throughout this book — inspired by fractals and the beauty of chaos — are sure to delight.

Potential readers of this book have probably heard about fractals and chaos if they are reading this foreword. These days computer-generated fractal patterns are everywhere. From squiggly designs on computer art posters to illustrations in the most serious of physics journals, interest continues to grow among scientists and, rather surprisingly, artists and designers. The word "fractal" was coined in 1975 by mathematician Benoit Mandelbrot to describe an intricate-looking set of curves, many of which were never seen before the advent of computers with their ability to quickly perform massive calculations. Fractals often exhibit self-similarity which means that various copies of an object can be found in the original object at smaller size scales. The detail continues for many magnifications — like an endless nesting of Russian dolls within dolls. Some of these shapes exist only in abstract geometric space, but others can be used as models for complex natural objects such as coastlines and blood vessel branching. Interestingly, fractals provide a useful framework for understanding chaotic processes and for performing image compression. The dazzling computer-generated images can be intoxicating, motivating students' interest in math more than any other mathematical discovery in the last century. In this book, Chapman and Sprott use poetry and computer graphics to form a mental lens that reveals beautiful forms in geometrical and cognitive spaces.

Physicists are interested in fractals because the practical side of fractals is that they can sometimes describe the chaotic behavior of real-world things such as planetary motion, fluid flow, diffusion of drugs, the behavior of inter-industry relationships, and the vibration of airplane wings. Often, chaotic behavior produces fractal patterns. Traditionally when physicists or mathematicians saw complicated results, they often looked for complicated causes. In contrast, many of the shapes in this book exhibit the fantastically complicated behavior of the simplest formulas. The results should be of interest to artists and non-mathematicians, and anyone with imagination and a little computer programming skill.

A number of the colorful patterns that follow result from chaotic attractors and intricate dynamical systems. I'm sure that Dr. Sprott's computer graphics experiments in these areas were inspired by a number of researchers' work over the past few decades. The early days of modern dynamics span half a century, starting with French mathematician Jules Henri Poincaré and Russian mathematician Alexander Mikhailovich Lyapunov in the early 1900s. The geometry of chaotic dynamical systems was firmly established in 1892 by Poincaré in his research on celestial mechanics. George Birkhoff's 1932 paper on remarkable curves was one of the first papers in the mathematical literature to discuss chaotic attractors. After undergoing a relatively quiet period, the study of dynamics was revived in the 1960's, partly as a result of the increasing use of computer graphic strategies of doing mathematics. In fact, digital simulations of the properties of periodic motions in nonlinear oscillations have become increasingly important, especially since chaos in real physical systems was clearly demonstrated in 1963 by MIT scientist Edward Lorenz, an atmospheric scientist who proposed a simple model for atmospheric convection that displayed unpredictable behavior.

Students and other readers fairly new to the chaos field may want to review some of the fundamental chaos concepts to better appreciate the images in this book. For example, readers should have some familiarity with dynamical systems that provide a deep reservoir for striking images. Dynamical systems are models containing the rules describing the way some quantity undergoes a change through time. For example, the motion of planets about the sun can be modeled as a dynamical system in which the planets move according to Newton's laws. Generally, portraits of dynamical systems track the behavior of mathematical expressions called *differential equations*. Think of a differential equation as a machine that takes in values for all the variables and then generates the new values at some later time. Just as one can track the path of a jet by the smoke path it leaves behind, computer graphics provide a way to follow paths of particles whose motion is determined by simple differential equations. The practical side of dynamical systems is that they can sometimes be used to describe the behavior of real-world things such as planetary motion, fluid flow, and engine vibration.

Opinions have always been mixed about whether fractal computer graphics are art. What do you think? Let the authors know your feelings. If Henri Matisse or Joan Miro were alive today, would they forsake their canvases and brushes for a computer terminal? Would they experiment with fractals and chaos? With computers, artists can collaborate with their colleagues over the Internet, as I have done on numerous occasions with Dr. Sprott. Perhaps Matisse and Miro would spend time inventing entirely new computer input devices to substitute for today's mouse. These devices would allow them to precisely emulate their own masterful brush stokes or go far beyond brushes into new territories many of us non-artists can barely imagine.

Computers will no doubt continue to facilitate serious future discoveries in the area of fractals and chaos. Interestingly, the fractals in this book are examples of gorgeous shapes created by very simple formulas. The Mandelbrot set, the icon of the fractal in the 21st century, is a striking example of a simple formula that gives rise to infinite detail and beauty. Arthur C. Clarke in *The Ghost from the Grand Banks* once noted:

> "In principle ... [the Mandelbrot set] could have been discovered as soon as men learned to count. But even if they never grew tired, and never made a mistake, all the human beings who have ever existed would not have sufficed to do the elementary arithmetic required to produce a Mandelbrot set of quite modest magnification."

Dr. Mandelbrot himself discussed his discovery of the set in a 2004 *New Scientist* interview:

> "Its astounding complication was completely out of proportion with what I was expecting. Here is the curious thing: the night I saw the set, it was just wild. The second night, I became used to it. After a few nights, I became familiar with it. It was as if somehow I had seen it before. Of course, I hadn't. No one had seen it. No one had described it. The fact that a certain aspect of its mathematical nature remains mysterious, despite hundreds of brilliant people working on it, is the icing on the cake to me."

I hope this book becomes the icing on the cake for *you* as you ponder the mysteries of mathematics, art, and poetry — and educate yourself with this stunning introduction to a field with mammoth repercussions in science and art.

Cliff Pickover, Ph.D.
www.pickover.com
Author of *A Passion for Mathematics*

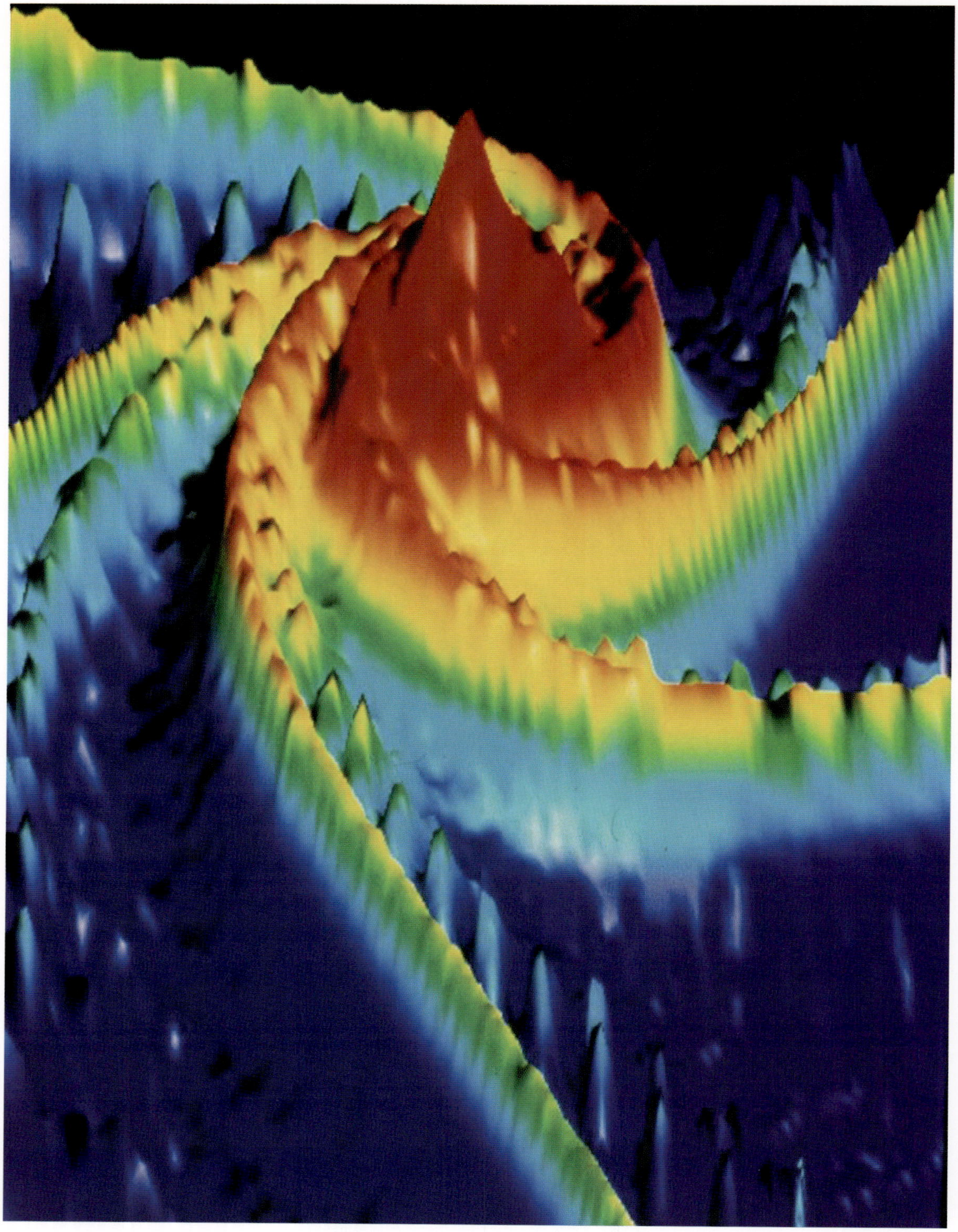

Mount Fractalia, a 3-D rendering of a generalized Julia set from *Keys to Infinity* by Clifford A. Pickover based on a collaboration with J. C. Sprott.

Preface

This book grew in part out of a weekly interdisciplinary Chaos and Complex Systems Seminar[1] that was founded in 1993 at the University of Wisconsin–Madison and that we have nourished along over the past decade. The seminar draws an eclectic audience of which we are but two examples, representing an unlikely collaboration between a plasma physicist (JCS) and a child language researcher (RC). Over the years we have listened to colleagues from Madison and elsewhere speak on topics as diverse as the nature of consciousness, robotic chemical analysis, black holes, the epigenetic control of spots on the buckeye butterfly, the dynamics of love and happiness, predator-prey relationships, chaos in plasmas, policy effects on the health care system, chaotic compositions for string quartets, cellular automata and the Game of Life, strange attractors, the dynamic control of millipede walking and healthy hearts, the economics of currency exchanges, weather prediction in a time of global warming, the evolving landscape, child language development, and much more. You will encounter many of these topics in the poems here.

Our individual interests led to a textbook on chaos[2] and work on a dynamic systems approach to child language development.[3] It also led us into work on the art that arises from depictions of complex system dynamics in space and in language. The images and poems of a complex world are the subject of this book, meant for the reader interested in the art, poetry, and the ideas of chaos. This is a book for browsing, for picking up and putting down, for clarifying the mathematician's use of a term, even for testing one's comprehension, for translating ideas into daily life, and seeing the complex way our lives evolve. It is for finding images that you enjoy and poems that speak to you. It is for appreciating the great diversity of artistic patterns that a few simple equations or rules can engender and for offering both metaphorical and mathematical ways to think about dynamical systems concepts through poetry and more formal definitions. We have tried to honor both logic and intuition, both eye and ear, and the different ways of knowing and perceiving.

The art here comes from simple equations, summarized in the appendix, and programs investigating millions of versions of these equations with changes in their parameters.[4] Through human artist tutors, one of us (JCS) taught the computer to choose the patterns most likely to

[1] http://sprott.physics.wisc.edu/Chaos-Complexity/

[2] J. C. Sprott, *Chaos and Time-Series Analysis*, Oxford University Press: Oxford (2003).

[3] R. S. Chapman, Children's language learning: An interactionist perspective. *Journal of Child Psychology and Psychiatry* 41, 33–54 (2000).

[4] J. C. Sprott, *Strange Attractors: Creating Patterns in Chaos*, M&T Books: New York (1993).

appeal to the human eye, and printed them in vivid colors. Out of years of computer search, producing hundreds of thousands of examples, we have chosen the patterns included here, instances of strange attractors, Julia sets, and iterated function systems (of which, more in the book!) that pleasingly portray the fingerprints of chaos.

The other one of us (RC) has composed poems that come from attention to changes in the natural world and human experience, and, in many cases, from seminar topics. Poems themselves are dynamic systems of sound and meaning, needing surprises for the ear, heart, and mind as they evolve through the reader's experience. These are poems that debate ideas — free will versus determinism, finite versus infinite, and join them to the topics that are poetry's themes: life, love, death, and nature. They are poems of science and philosophy; but also of children's games, family lost and gained, and our only Earth.

We have put together in these pages ideas, words, and images that have delighted, surprised, puzzled, entertained, and educated us. We hope you will find your own favorites. You can see us or view our work in detail at our Web sites.[5]

Robin Chapman
Julien Clinton Sprott
Madison, Wisconsin
March, 2005

[5] http://sprott.physics.wisc.edu/ (JCS),
http://ww.madpoetry.org/madpoets/chapmanr.html, and
http://www.comdis.wisc.edu/facstaff/rchapman/index.htm (RC)

Acknowledgments (Robin Chapman)

My thanks go to the Leighton Studios, Banff Centre for the Arts, Canada, where many of these poems were written and the book assembled; to the Wisconsin Arts Board for Individual Artist Development Grants that supported some of the work; to the Chaos and Complex Systems Seminar at the University of Wisconsin–Madison for its many speakers on dynamic topics; and to my colleague Clint Sprott who taught me what I know about chaos and nonlinear dynamic systems. My thanks, also, to my manuscript groups, for their patient comments; and to the journals in which some of these poems first appeared. We are grateful to our chaos seminar colleagues Derek Bownds, Myrna Casebolt, and Russell Gardner, Jr. for suggestions, and to Susan Elbe, poet, and Monica Schokkenbroek, graphics artist and photographer from Amsterdam, NL, for their advice on the manuscript.

Acknowledgments (Clint Sprott)

My thanks go to my colleagues in the UW–Madison Department of Physics who have tolerated my frequent excursions into subjects that are not conventional physics but that have given me so much pleasure and satisfaction. The late Donald Kerst (1911–1993) launched my career in science and nurtured me like a father. George Rowlands introduced me to chaos many years ago and continues to be a valued collaborator and mentor. Ted Pope convinced me that these images, which I first produced quite by accident during my mathematical investigations, have artistic merit. Cliff Pickover provided an inspirational example through his creative thinking and boundless energy. I am grateful for the interactions with my colleagues at the Chaos and Complex Systems Seminar. My coauthor, Robin Chapman, has been a constant source of inspiration and encouragement for over a decade, and I greatly admire her ability to visualize and verbalize complex concepts, even if she has never quite convinced me that a poem is more beautiful than an equation.

Acknowledgments of Previously Published Poems (RC)

5 AM: "Missing Cam."

Appalachia: "The Avalanche Lilies of Paradise," "Dear Ones, 12/12, Madison," "Dear Ones, New Year's Eve, Banff."

Byline Magazine: "The Poets Read for Three Hours in the Grape Arbor."

The Christian Science Monitor: "Migration," (as "Assembly"), "The Dog Gets Loose," "Every Lake a Loon, an Eagle," "First Frost," "The Gardener," "The Gatherers of Stories," "November Canoeing," "Sand Barren," "Trees Full of Wind," "Walking to Work," "What We Don't Notice in Winter."

The Comstock Review: "Northern Pike."

A Cup of Poetry and a Side of Prose: "Stockpiling at Y-12." (also posted on http://poetsagainstthewar.org)

Cutbank: "I Join the Women of Churchill College."

Higher Ground Poetry Anthology: "Linear Systems."

Kalliope: "Ark for the 22nd Century"

Mad Folk News: "Summer Contra."

Nimrod: "Eospaltria australis: Dawn Singer."

Northeast: "Arriving," "Going Forth," "In the Time That is Left," "One Line of Clouds on the Horizon," "When You Fall into Dreamless Sleep."

OnEarth: "Hibernation."

Poetry Motel: "Fields."

Rosebud: "The Good Death," "Lying in Bed Flipping the Remote Control Through Thirty Channels," "Train Wreck."

Southern Indiana Review: "Time, Now."

The Sow's Ear Poetry Review: "Dear Ones, 12/13, Madison."

White Pelican Review: "You Leave and I Miss You."

Wisconsin Academy Review: "The Science of Life," "Suddenly the Sumac."

Wisconsin Poets' Calendar: "Before the Dog Saw a Squirrel," "Chickadees at Anvil Lake," "Flood," "Studying the Swallow's Flight," "Succession," "Dynamical Systems."

Wisconsin Trails: "Canoeing, Early April."

Words & Images: "The Volatile Wife."

Yankee: "Backyard."

"Ark for the 22nd Century" appears on the Future Website, Sprott's Gateway, http://sprott.physics.wisc.edu/predsamp.htm.

"Finding a Way" toured in the Epidemic Peace Imagery exhibit of works by over 160 poets and visual artists in Madison, Wisconsin and the state in 2003–2005.

"Strange Attractors" appeared in Sprott, J.C., *Strange Attractors: Creating Patterns in Chaos.* M&T Books, 1993.

"What's the Bear Doing in Your Poem?" and "Flight Cage, St. Louis Zoo," appeared in *Animal, Anima*, a composition for soprano, piano, trombone and tubular chimes by Sara Scott Turner, with texts by Rainier Maria Rilke and Robin Chapman. Commissioned by Ivory Echoes Chamber Music Series, University of Manitoba, Winnipeg, Canada.

Contents

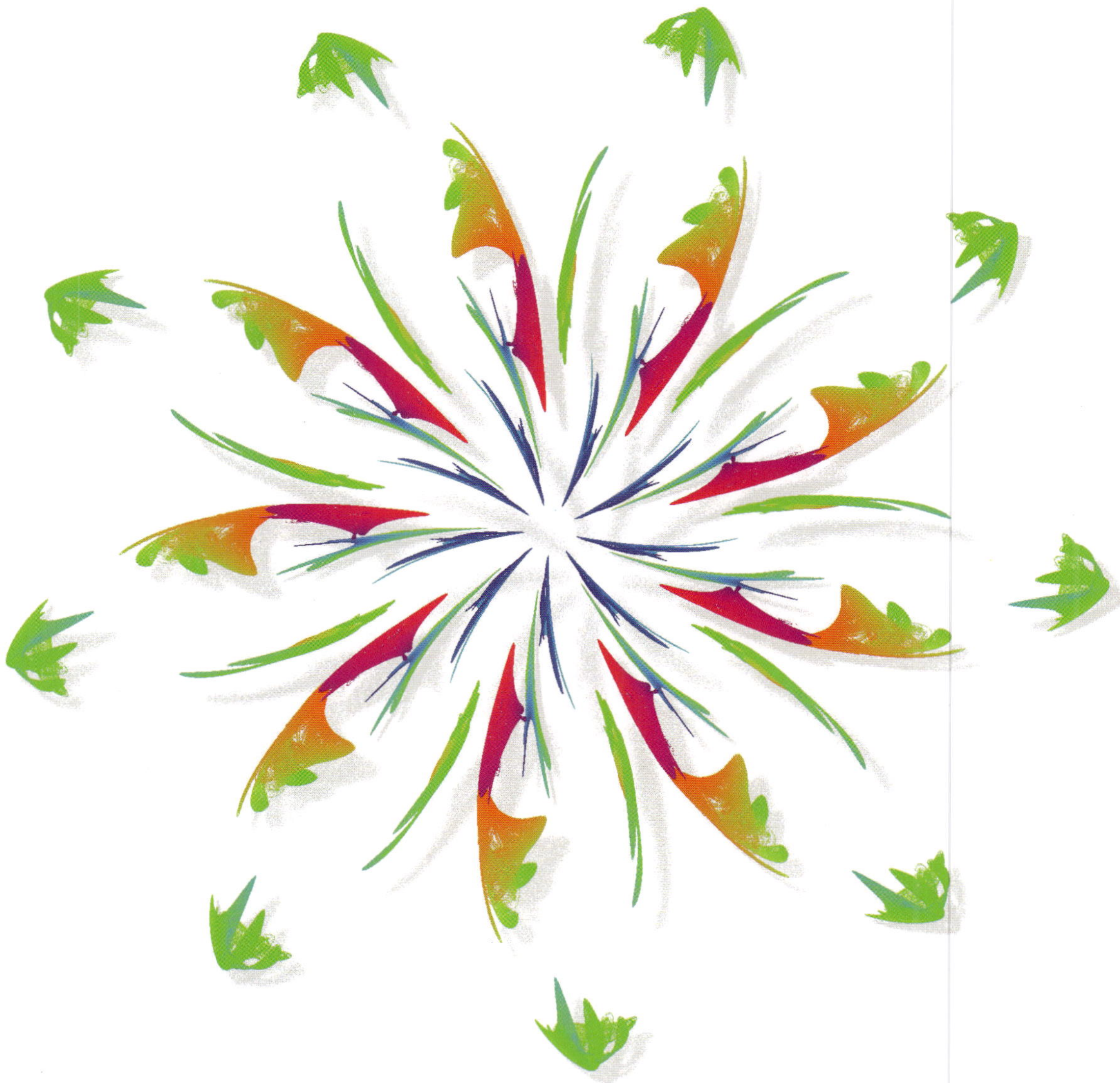

Chapter 1. Dynamical Systems

Everything flows, nothing remains. — Heraclitus

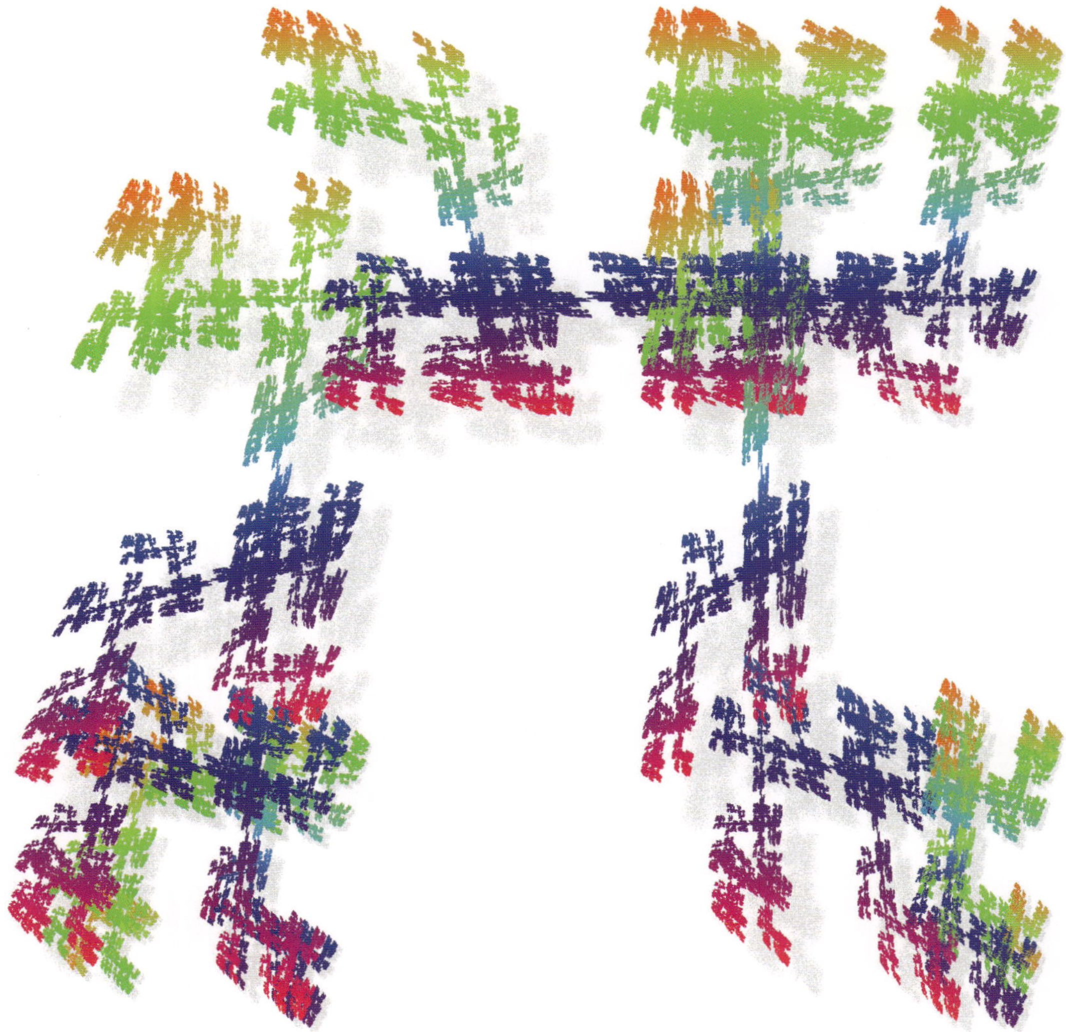

Dynamical system

A dynamical system is one that evolves in time according to some set of rules or equations. The rules may be deterministic, in which case the evolution is uniquely determined by the past history of the system, or random, in which case the evolution is governed by chance such as the flip of a coin at each time step. Think of an ant walking across a sheet of paper or a fly buzzing about the room. Although the rules may not be obvious, a spatial pattern results from plotting the motion. In these examples, the variables are the coordinates representing the constantly changing position. More generally, the variables may be more abstract, such as temperature and humidity in a dynamical model of the atmosphere, or stock prices and exchange rates in a dynamical model of the economy, in which cases there may be many more variables than can be simultaneously displayed on a computer screen or sheet of paper.

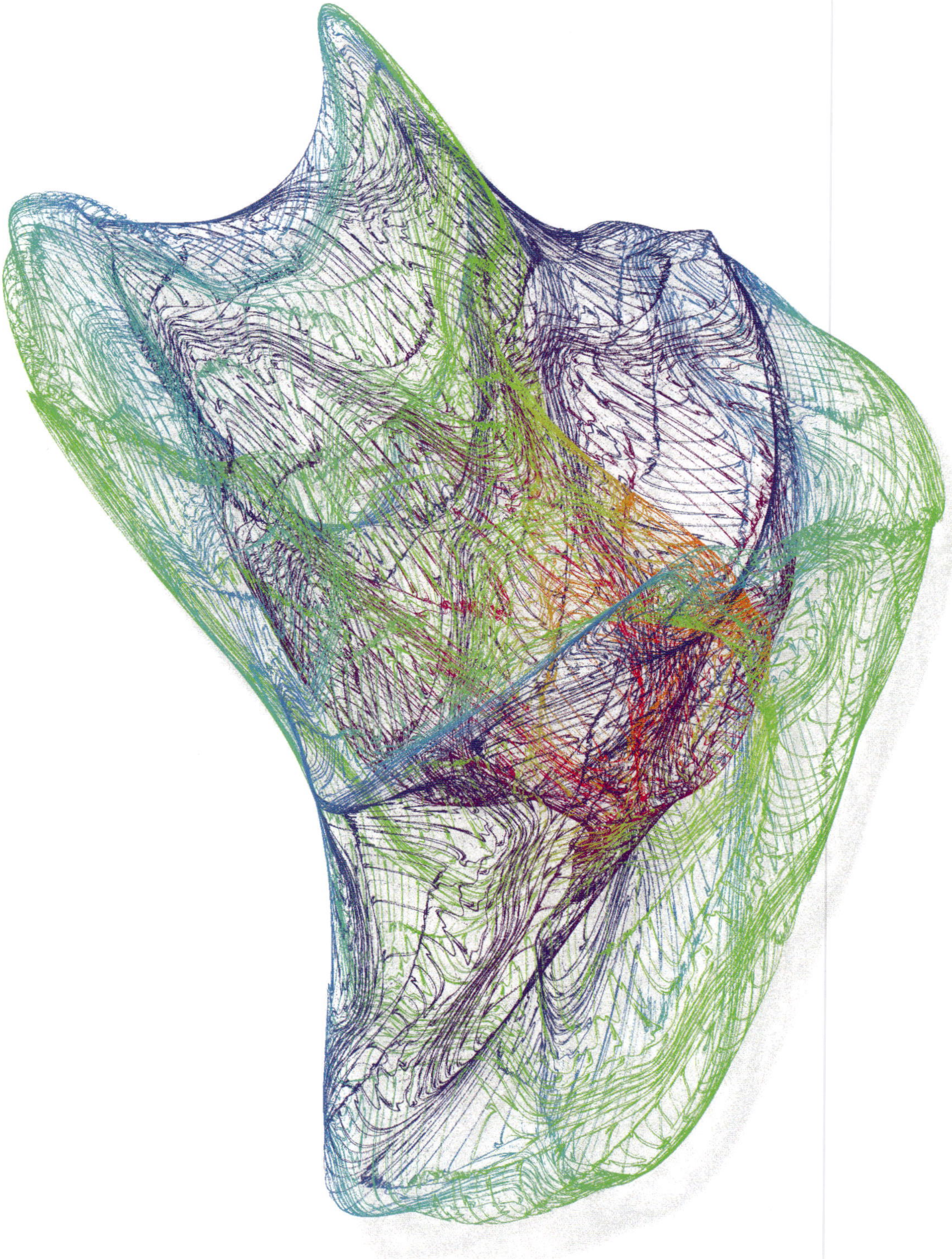

DYNAMICAL SYSTEMS

What's *not* changing in time?
The glass in the window pane
sags slowly, the sunlight
streams through the glass, the cat
washes her face with her paw,
the house gathers dust motes,
the geraniums we brought in
before frost take root and flower.
Outside, a wind is blowing the leaves about.
The universe we once thought steady-state
is flying apart. Inside, we are waltzing,
laughing, to the music of the nickelharppe.
And you, reader, anchored by gravity
and oxygen and eye, thinking now
of the sky full of stars, dancing, house repairs —
what strange, unpredictable pattern is yours?

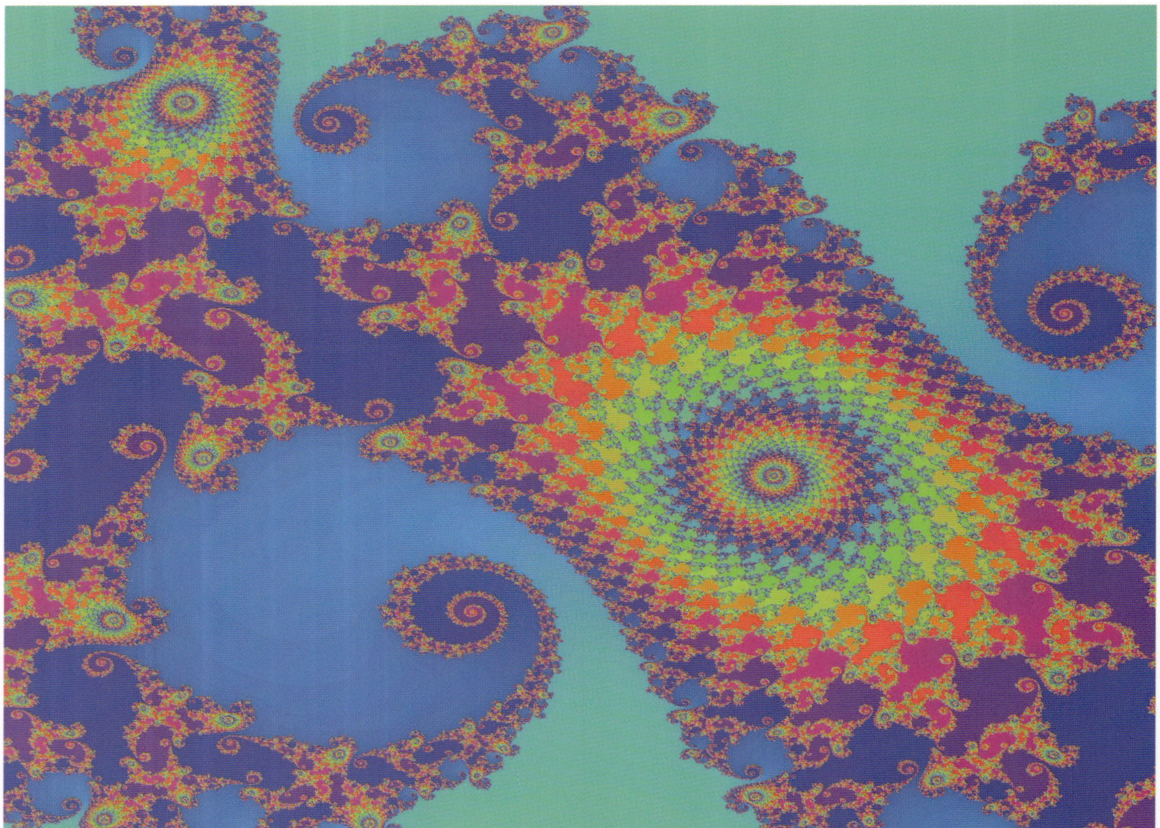

Simple or Complex?

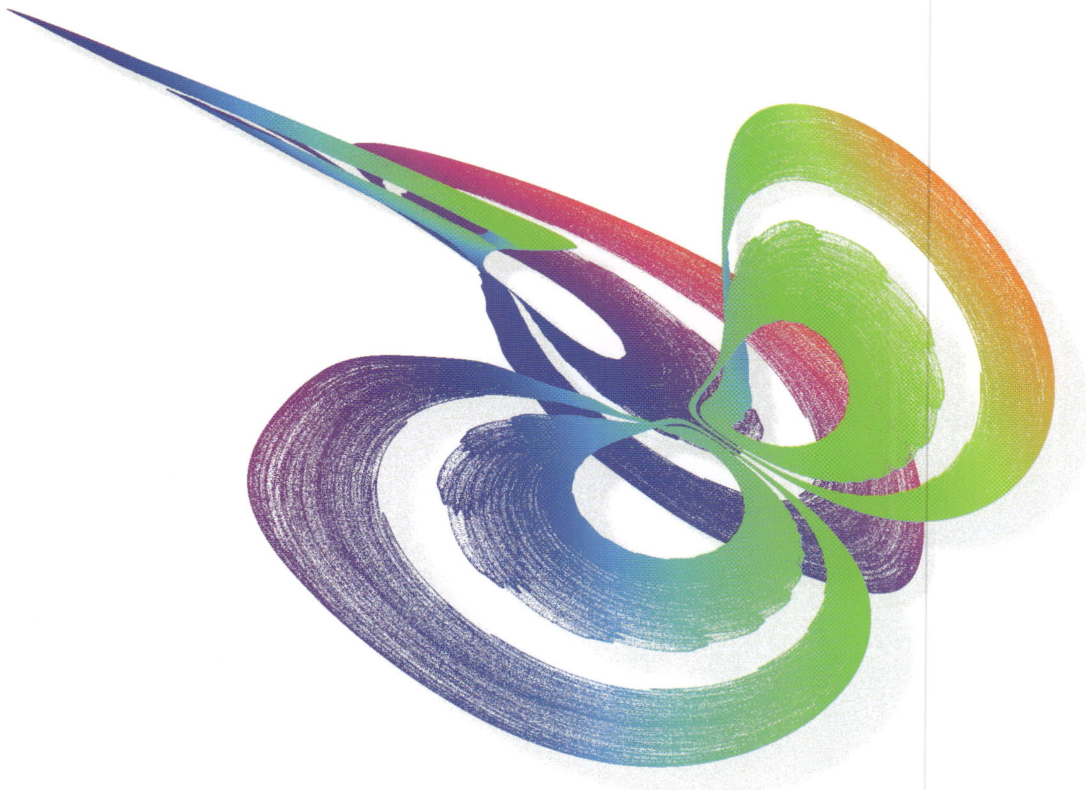

Complex system

A complex system is one that consists of many parts or variables that interact usually in a nonlinear way with one another. The rules describing the interactions can be very simple, and the interactions can be local (involving a small number of near neighbors) or global (involving all the parts of the system). Common examples include the ecology, the economy, the weather, the electrical power grid, the Internet, the World Wide Web, and the brain. You might expect the behavior of such systems to be complex, and it usually is, but such systems can also exhibit highly organized behavior. Even when started in a disordered initial state, they will often self-organize in apparent contradiction to the second law of thermodynamics, which states that the disorder (or entropy) in a system tends to increase. The explanation is that such systems are driven far from equilibrium by external forces, and they are able to lower their entropy by increasing the disorder of whatever is driving them. For example, the source of all the highly organized life on the Earth is the Sun, and the Sun is gradually cooling and coming to equilibrium with colder objects throughout the Universe.

THE GORILLA THAT WALKS THROUGH THE BASKETBALL GAME

Thirty professors at the chaos talk
on how the mind works,
we watch the white-shirted players
on the videoscreen, doing our assigned job
of counting the number of passes they make,
not so easy when two different balls
have appeared in play, and our counts
at the end of the videoclip vary —
eleven say some, fourteen insist others —
but we're feeling good
that we've kept our eyes on the balls
and the hands and the backs,
carried out our appointed task,
and when we're asked
if we noticed anything odd, no one nods;
though shown the replay
we see we've missed
the gorilla that wanders through
the twist of bodies,
crossing the court
from left to right in a leisurely way,
looking about with curiosity —
leaving us shaken by the query
of what else we've missed in our lives
keeping our eye on the ball.

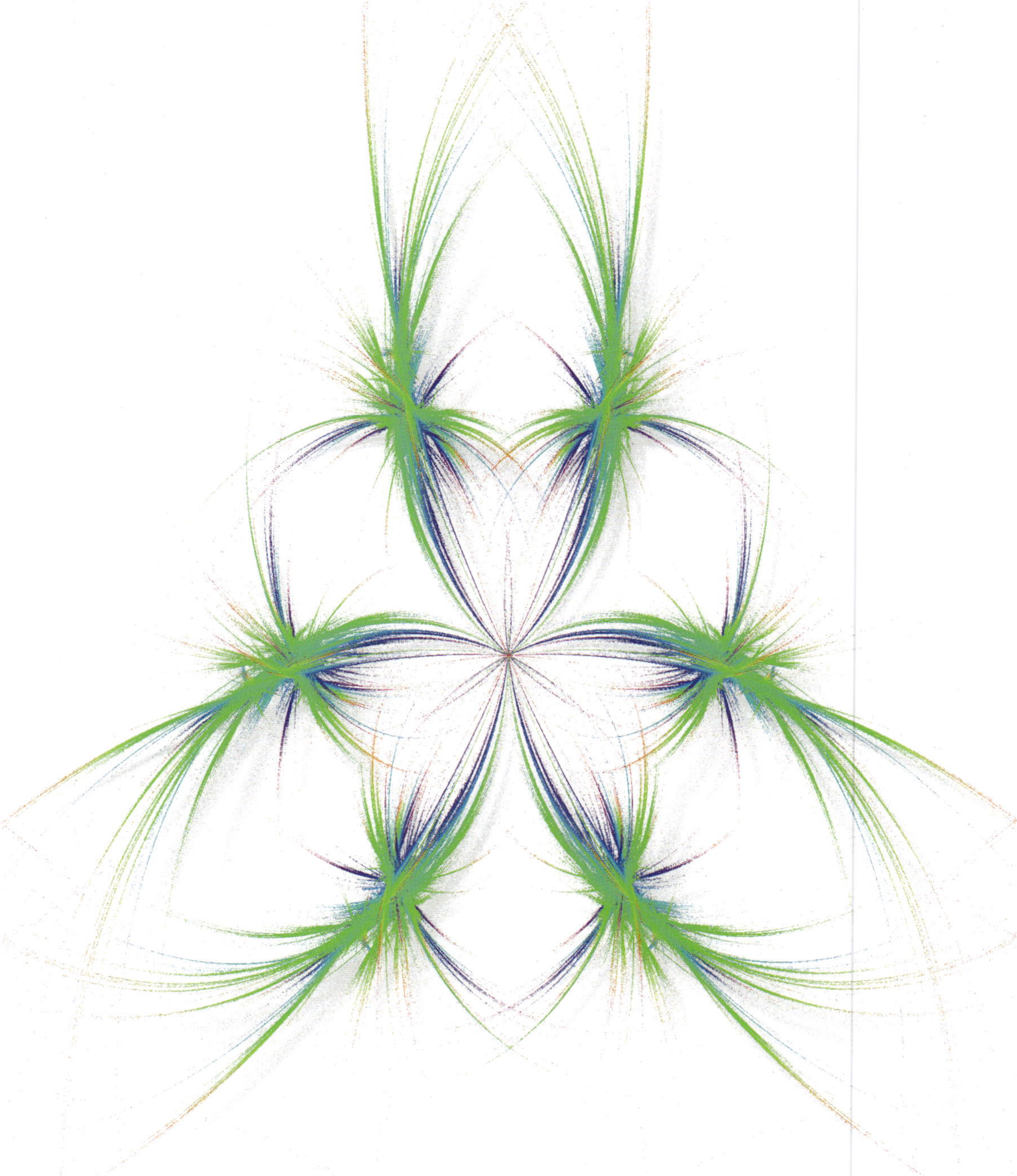

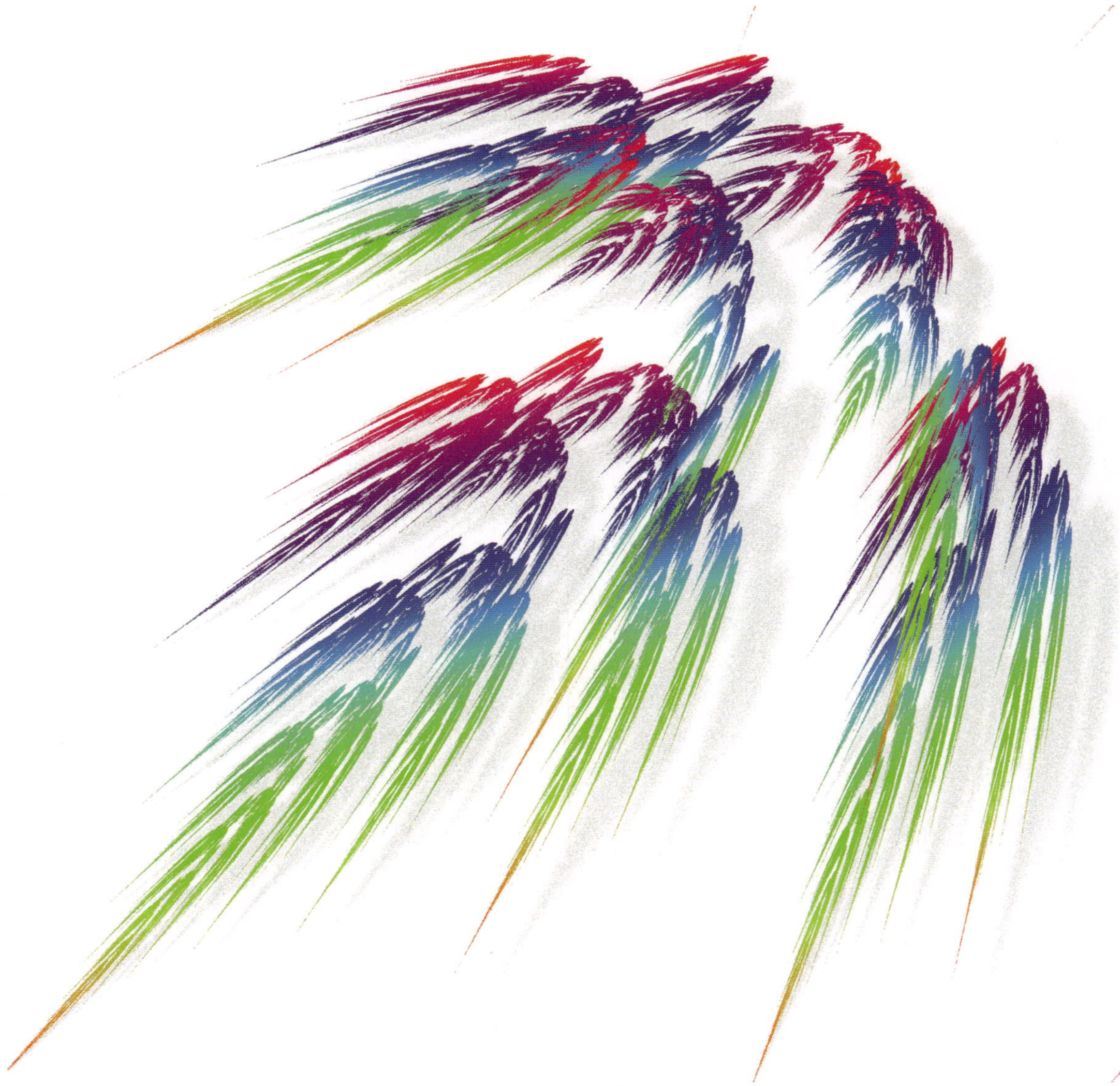

TREES FULL OF WIND

They are made for it,
Bending, sieving, slowing
The energy — a wild
Conducting, ushering
What passes invisibly
Into the sensual world —
Afterwards, deep roots
Tremble with the memory
Of dancing.

STUDYING THE SWALLOW'S FLIGHT

For years I've passed its evening notation
Of telephone lines, the mud and white splat
Of its home, and only now, watching it skim
The marsh, climb purple into the sky, veer, dive,
Brake for the willow branch, do I learn
How, opportunist of air, it moves and lands,
Fanning briefly its swallow tail to flash
On each spread feather the image of the sun —
Then, airborne, scoop through mosquito swarm.

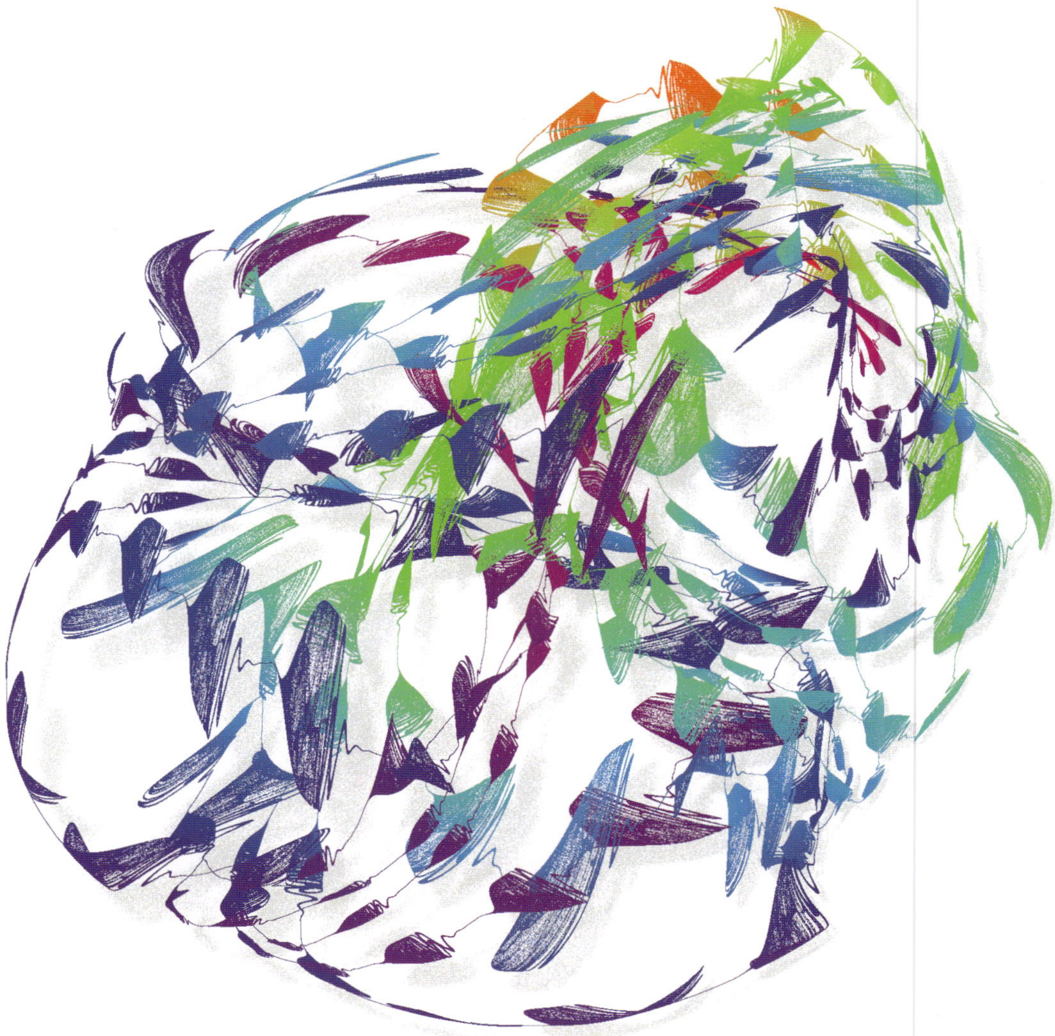

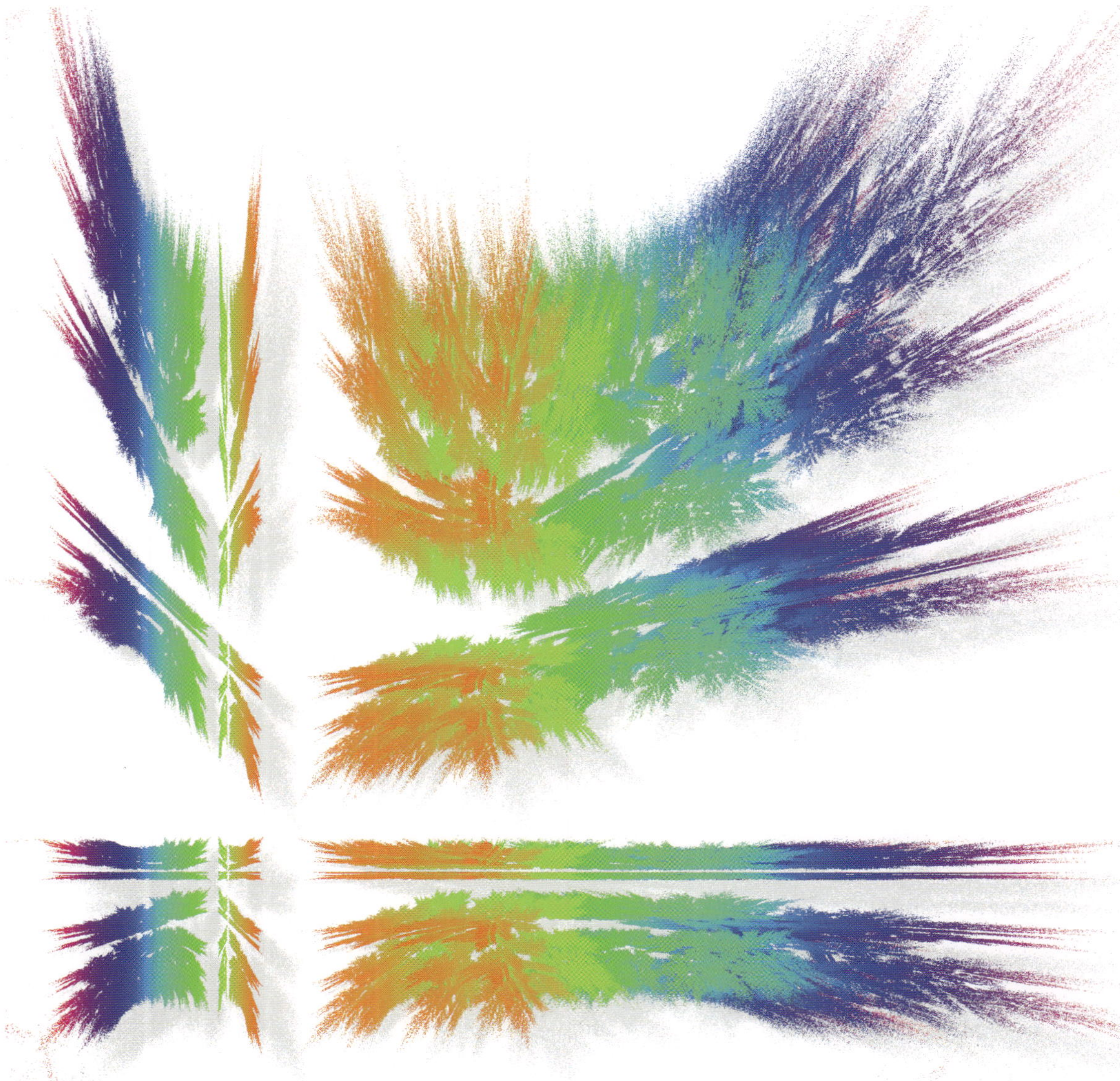

A MOMENT OF EXTREME COMPLEXITY

1978, a Surabayan roadside, ten lanes crowded with bicycles bearing lawnmowers and bunches of squawking chickens tied by their feet, becaks pedaling men and goats seated side by side, motorcycles trailing blue exhaust, boys and their girls astride, who balance chests of drawers, panes of window glass on their laps; water buffaloes pulling carts of mattresses, semitrailers loaded with oil, a child at the other side of the road throwing gasoline on the charcoal fire where sate is broiled — the becak driver I've hired to cycle me across the road holds up one hand and starts us across, looking neither right nor left on the endless straight line to the other side.

THE TRAVELING SALESMAN'S PROBLEM IS NP-DIFFICULT

***NP-difficult*: nondeterministic polynomial problem whose difficulty increases faster than its size. The salesman's problem is to schedule his visits to cities minimizing the distance traveled.**

We were all for optimization of student opportunities
for taking courses and minimization of teaching time
and thought it would be simple enough for the volunteer
engineer grads to write a program for extra credit
to solve our annual scheduling problems
matching up 30 staff and 90 students and 200 clients
if only the faculty stopped being pig-headed
about their favorite teaching times and allowed the process
to begin with what each student wanted to take and when
each client wanted to come and from that determine when
each class could meet so that no student had a conflict
with class or client, so who knew it was a problem NP-difficult,
as hard to schedule as the Traveling Salesman or a herd
of cats? The program was still running long after classes
began, the schedule made up on the usual grid by the department's
graduate secretary, and when the computer finally coughed up
the recommendation that my Com Dis seminar in language
disorders meet at 11 p.m. Saturday night and the class in voice
at 6 a.m. Sunday, we put an end to optimal algorithms
and let the students and clients like it or lump it.

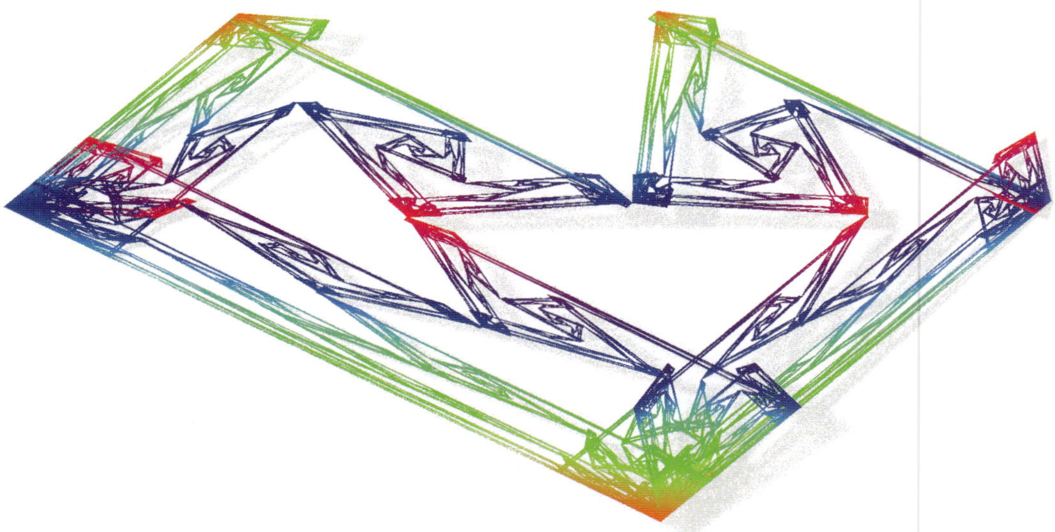

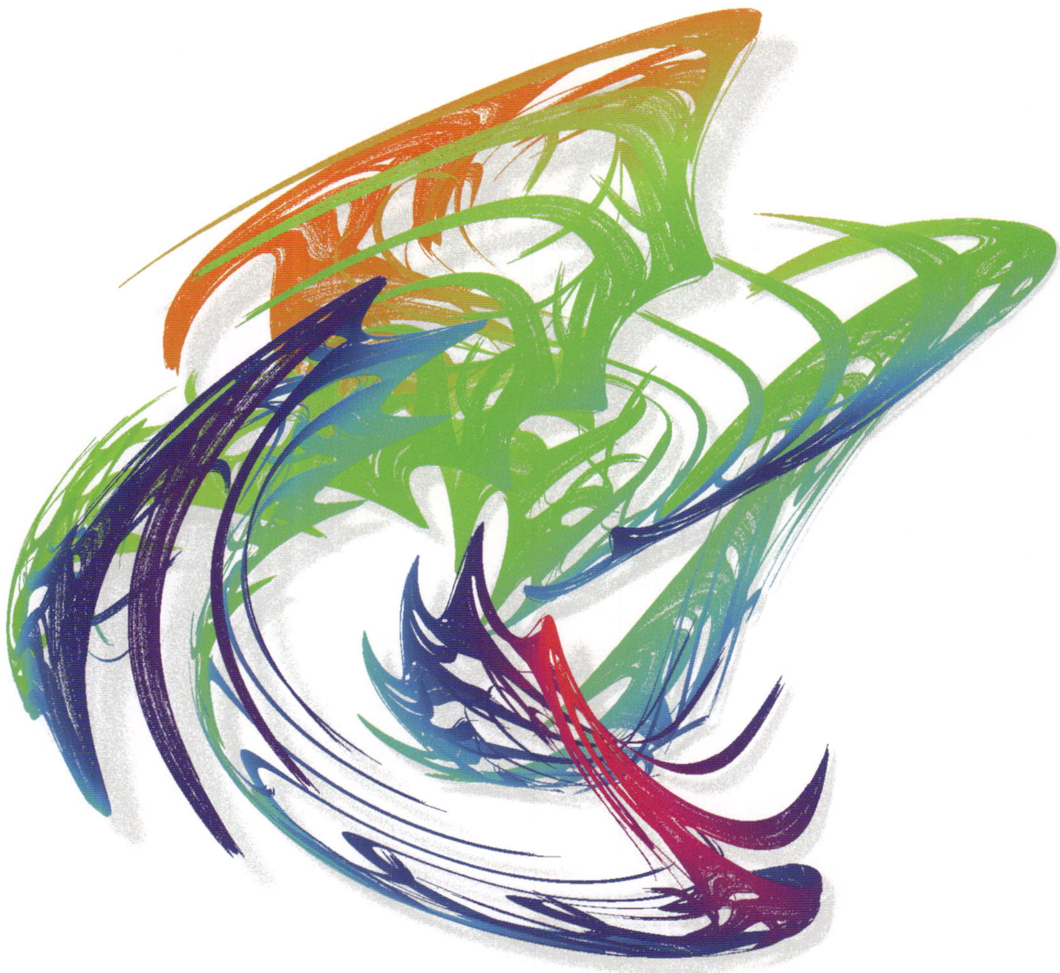

Linear or Nonlinear?

Linear system

In a linear system, the output is proportional to the input. Such systems also obey the principle of superposition, which states that if one input gives a certain output and a second input gives a different output, the two inputs combined give an output that is the sum of the individual outputs. Linear systems are described by linear equations, involving only sums and differences of the various terms, perhaps multiplied by constants. A linear system can contain many variables, and each variable can depend on all the others. A linear system is not one in which each effect has a single cause. Linear systems can exhibit quite complicated behaviors, including exponential growth and decay, periodic oscillations, and oscillations that are a combination of many different periods. Linear mathematical systems can always be solved in a straightforward way with predicable behavior. They can never exhibit chaos.

LINEAR SYSTEMS

We learned about them in school,
        outcomes predictable:
What goes up must come down.
        Every Jack has his Jill.
Tomorrow is another day
        in a line of days.
I thought this meant a straight line,
        plodding along a path
a child with a ruler had drawn,
        though mathematicians, lost
in their own circuitous thoughts,
        meant only that output
was proportional to input —
        you get what you pay for,
however accelerated the cost.

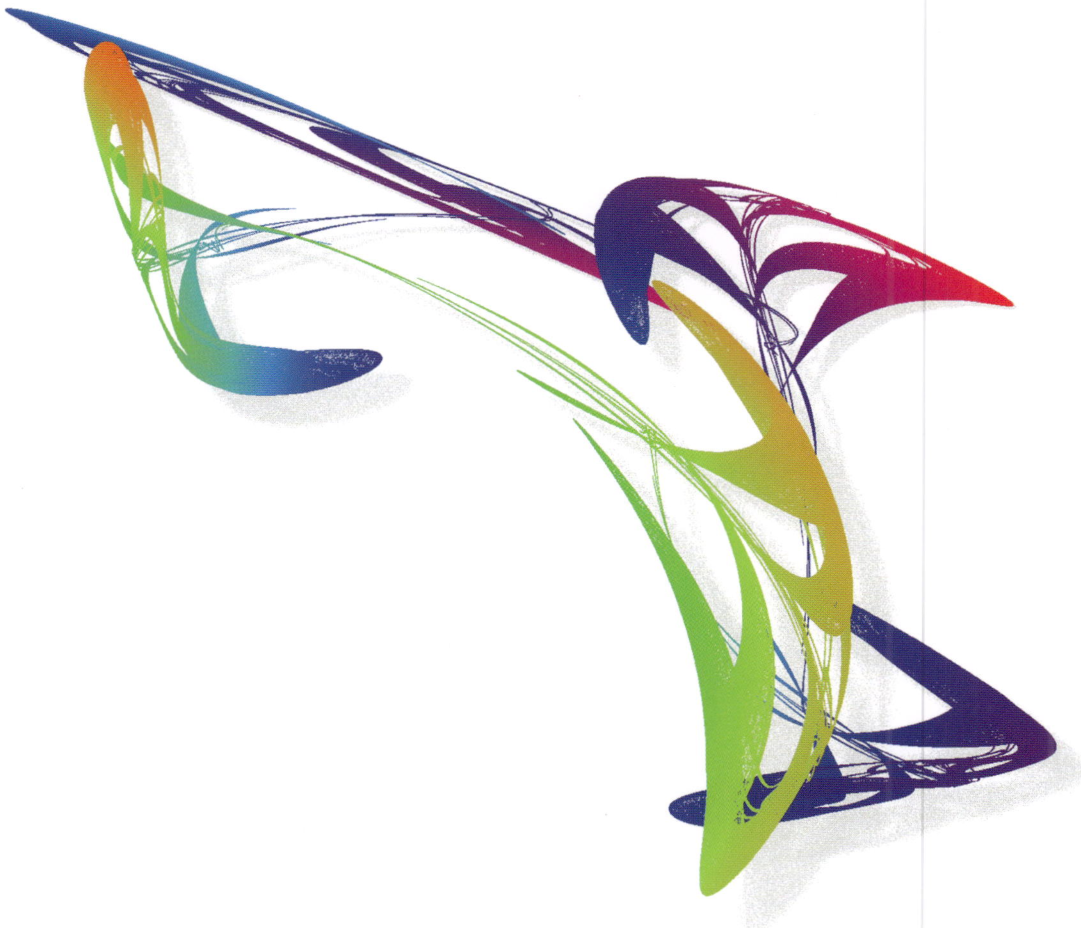

ARK FOR THE 22nd CENTURY

We'll ask them to line up two by two,
Our contradictions, opposites, contrary
Categories — all that we've fought
Over for the centuries —
What else to do but bequeath
Our own polarities?

Either shyly takes *or*'s hand,
Hard-headed finds *soft-hearted*,
Right lines up with *wrong*,
East with *West*, *North* with *South*,
Rich takes *poor*, *mind* admits *body*,
Decimation asks for *birth*.

We'll close the hatch and lock it,
Set them adrift in an acid rain
Long enough to hope
For children of hybrid vigor
From those contentious
Unions– polyglot *both*, *all*,
When the dove returns
With the olive branch
And the doors re-open
On our one precarious world.

Set of measure zero

A set of measure zero is a group of things that occur with zero probability. You might think this means that they never occur, but in fact they can occur infinitely often. For example, a line (in the mathematical sense of having length but no width) has no area when drawn on a sheet of paper, and yet infinitely many points make up the line. The points are a set of measure zero in the plane since you could throw darts at the plane forever and never hit one of them (assuming the dart is infinitely sharp). A strange attractor is also a set of measure zero in the state space of its variables.

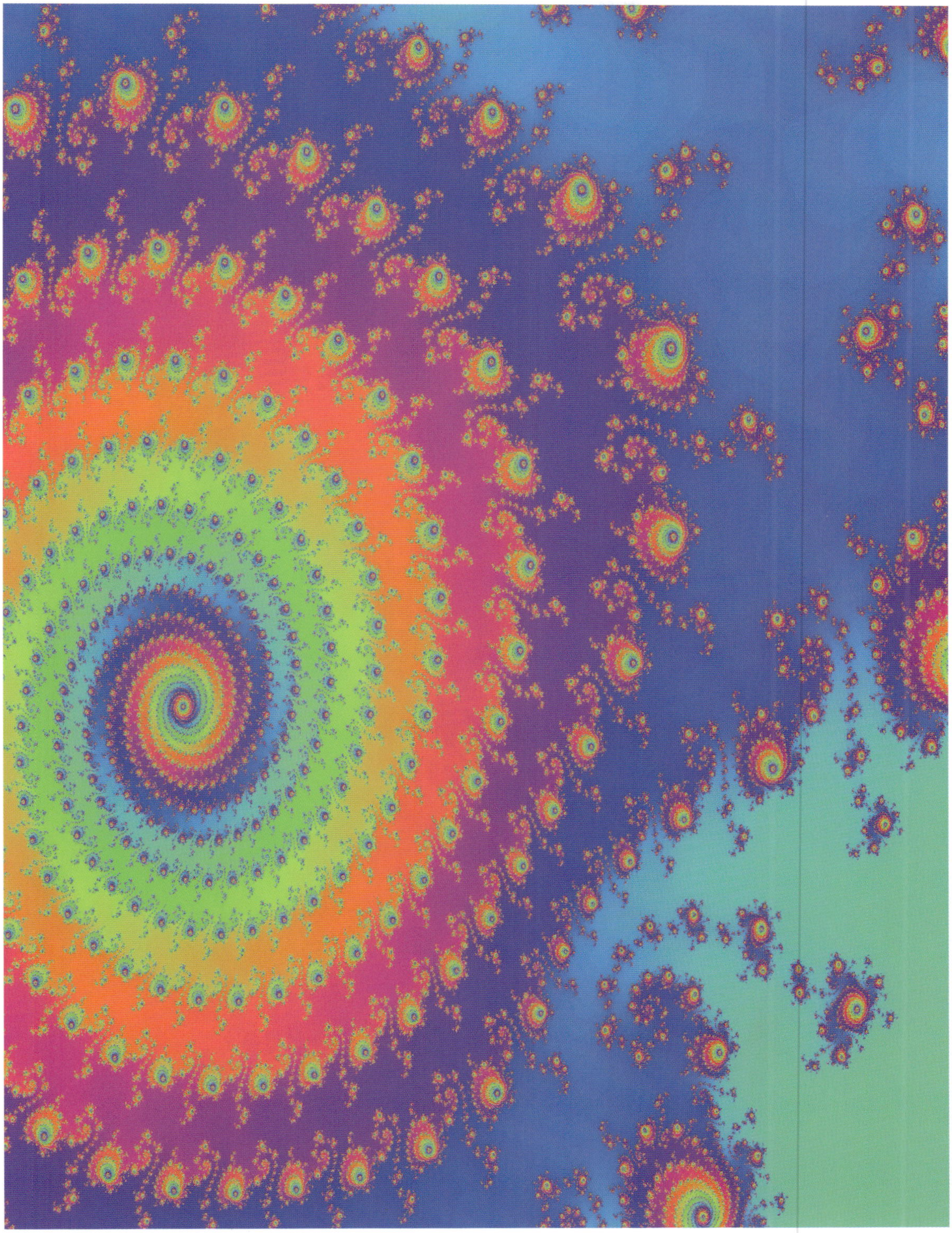

THE LINE

The path, always moving, of the train
Hauling freight through the night.
The bicyclist following, the next century.
The meditator, learning as she walks
The space between one step and the next.
And those corpuscles traveling our bloodstream,
Bearing in each arterial circuit new breath to new work.
The nerves of the hand reporting the static
Of the tactile world, its roughness and slick,
Heat and cool, flickering our brain and gut —
Lines connected, branching out, whole fishing nets.
Not the line of one-dimension that mathematicians dream:
No crossings, no knots, just this point, and the next,
And the next — iterated, with no start, no stop.

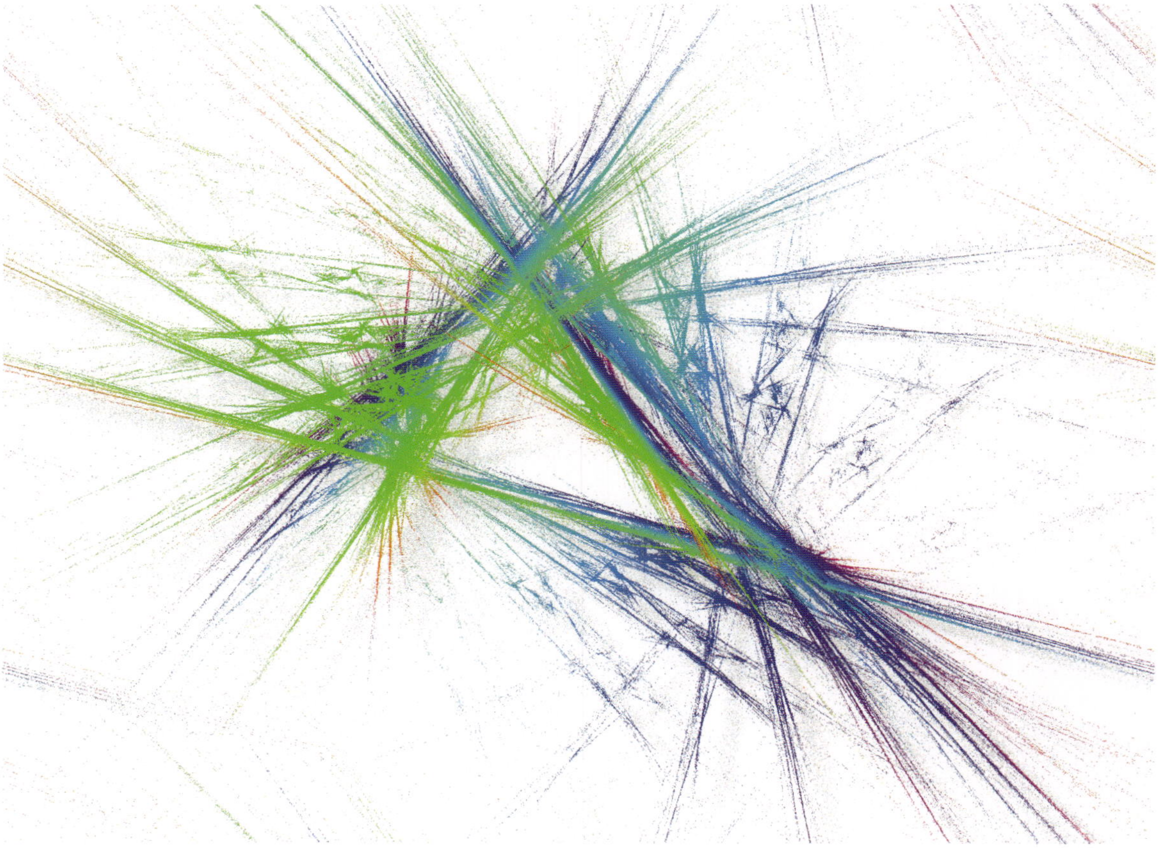

Nonlinear system

A nonlinear system is one in which the output is not proportional to the input or for which the whole is not equal to the sum of its parts. In a mathematical system, a nonlinearity is recognized by a term that involves a product or ratio of two or more variables such as XY or X/Y, or a variable multiplied by itself such as X^2, or perhaps something more complicated such as an exponential function or a trigonometric sine or cosine. Deterministic dynamical systems need a nonlinearity to produce chaos. By contrast, the solution of linear deterministic dynamical systems can only grow or decay exponentially, oscillate periodically, or remain fixed. Linear systems can be broken into parts, whose behavior can be predicted, and then put back together by simply adding up the behavior of the various parts. Nonlinear systems are usually much more complicated, and their behavior usually cannot be predicted without the benefit of computer simulation.

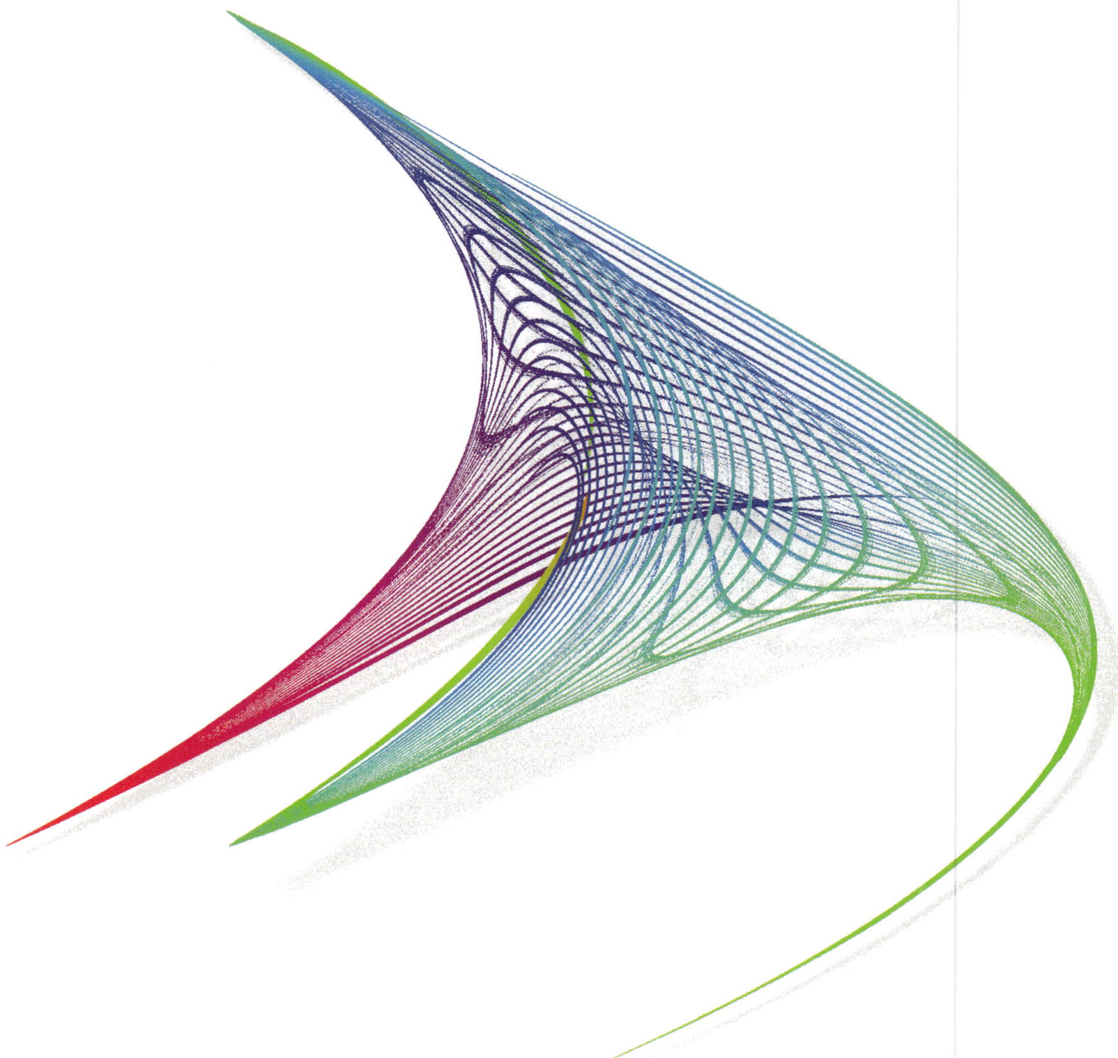

NONLINEAR FUNCTION

Def. 1: Not a linear function

A nonlinear life doesn't turn out
The way you might expect,
More of the same each year,
Straight line from birth to death —
But more like David Copperfield's
Or Horatio Alger's, the story
Of any infant
Where a look or word
Made all the difference
In who they became today.

Def. 2: One in which $f(x+y)$ does not equal $f(x) + f(y)$

This is easily enough understood
By any child of divorce — Mom's house
And Dad's house are not the same
As the house with both Mom and Dad before.
Or think of f as happiness,
And know that what they had together
Is not what they have now, whatever
The plus or minus sign of once-upon-a time.

Def 3: One in which $f(ax)$ does not equal $af(x)$

This one's obvious to poets, mail carriers,
Preschool teachers — a salary of a for each x,
Equal pay, is not the same as 500a for the CEO
And what's left over for the rest of us $x$es, even though
The nation's average annual income would remain
The same, and some economists would claim
Measures of average income are perfectly adequate
In charting progress in the economy's name.

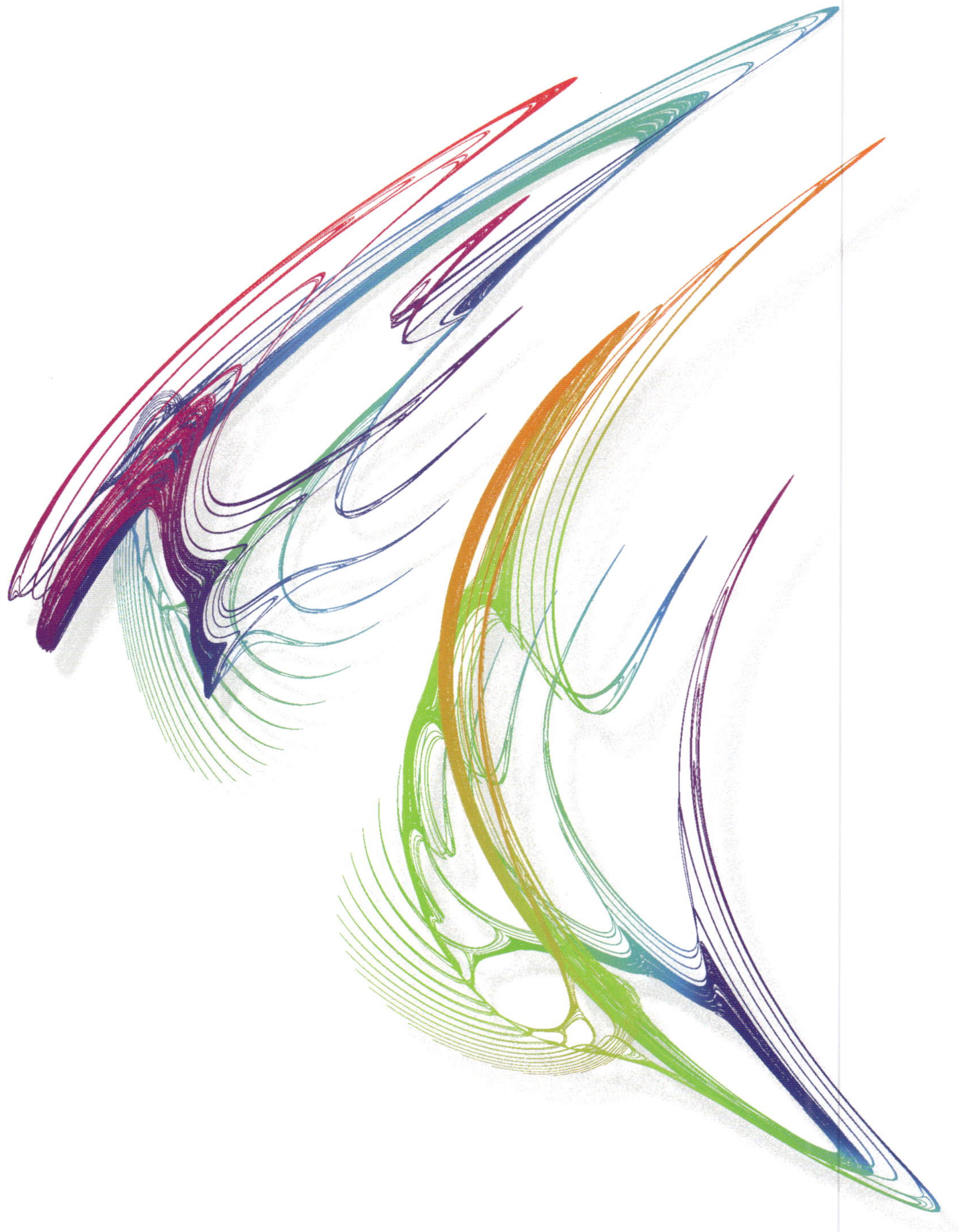

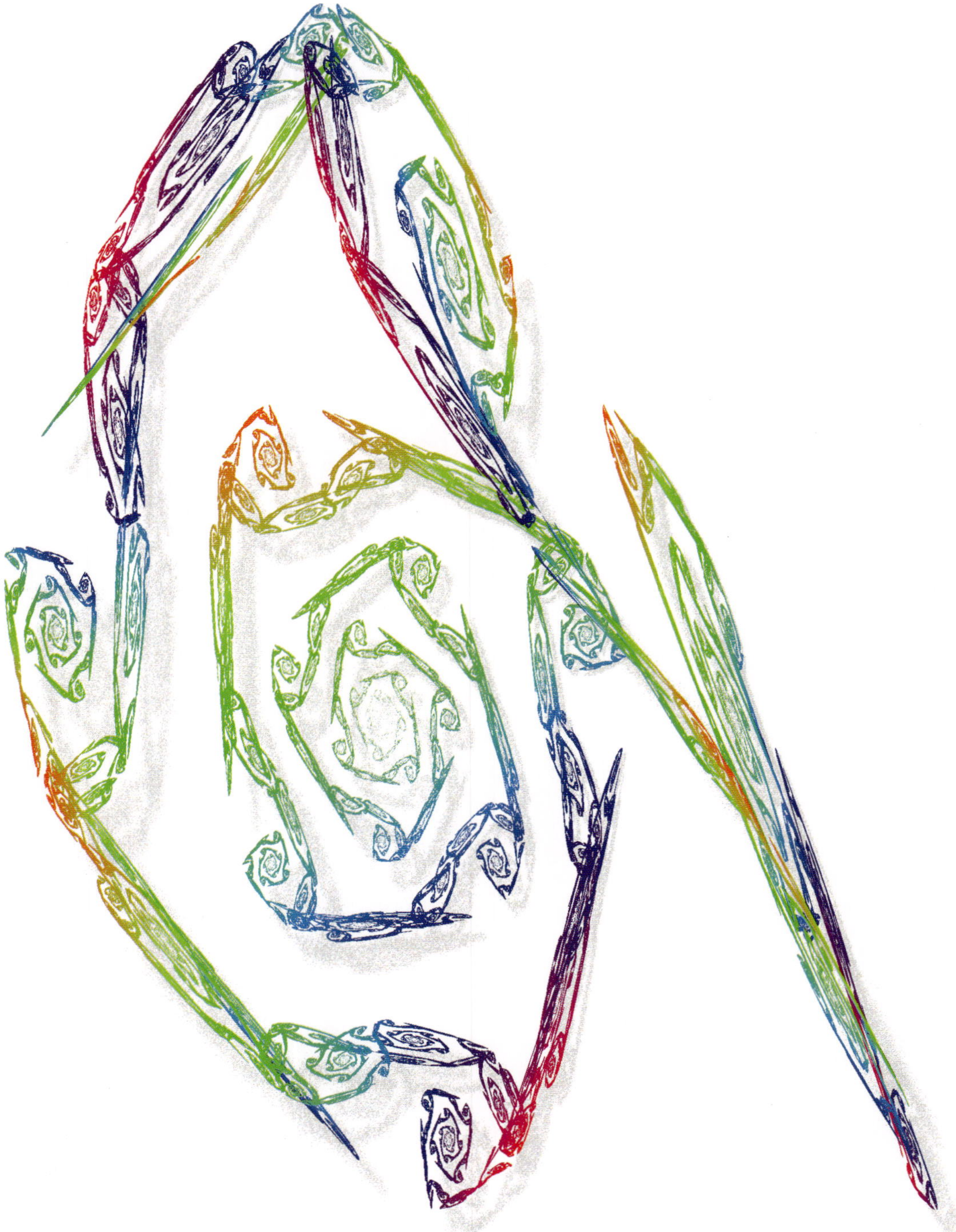

GRAIN OF SAND

for John Valley

Our speaker shows us the photograph,
        a single grain of worn zircon the width
of two human hairs — the oldest
        piece of rock in the world, found
in the Jack Hills of Western Australia
        after a search around the earth.
He's used an ion-probing machine the size of a house,
        working days and nights to infer
its age and temperature at the start. We stare
        at the rubbed edges of black crystal,
its facets cookie-cut
        by the sampling probe, try to imagine
its years: 4.4 billion of our 4.5, the data say,
        and arising in times cool enough for water,
meaning 1.5 million years more time for early life
        to arise than prevailing theories assume —
sand trailing its history of oceans and mountains,
        the mix of gases and lightning, the brew
of living soup, the imprint of late lunar cataclysm
        when meteors plunged to earth's core,
tore off the moon, and life pulled through in a cell
        surviving in some deep sea hot-vent. The mind
starts remaking the story of how we began —
        star people falling through a hole in the sky,
carried on Turtle's back, arising from the dung heap
        of history — peering again into Blake's grain of sand.

Three-body problem

Whereas, the "two-body problem," such as a planet orbiting the Sun was solved about 400 years ago by Johannes Kepler (1571–1630), the same problem with three bodies such as the Earth, Moon, and Sun, has never been exactly solved, and probably never will be. The reason is that the motion of three or more astronomical bodies, all interacting with one another, is usually chaotic, and chaotic systems cannot be predicted arbitrarily far into the future because of their sensitive dependence on initial conditions. We can predict the motion of the Earth and Moon very accurately for hundreds or even thousands of years, but these predictions eventually fail after millions or billions of years. Whether the Solar System is stable or not is unknown. It has existed for billions of years, but there may have been additional planets that long ago were ejected from the Solar System, and it is possible that some of the existing planets, even the Earth, might some day also be ejected or bumped into a very different orbit around the Sun.

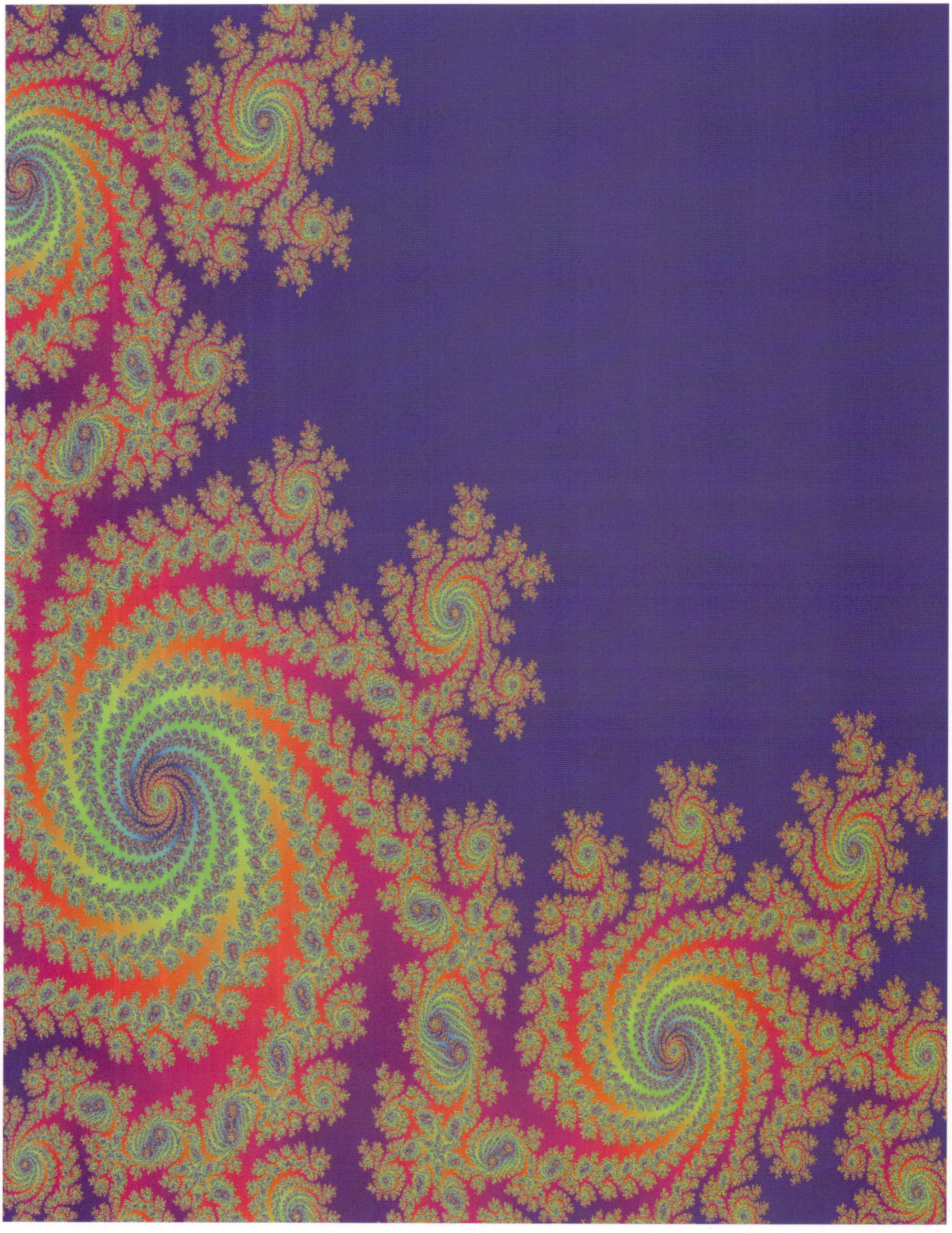

CLOCKWORK

for John Mathis

At the summer Sunday service
the learned astronomer describes
the explosion of knowledge in his field,
news spewed back by satellite
of a nearby black hole, sucking
in matter, jetting out gas
only 50 million light years away —
if our own sun collapsed so
under the crush of gravity, we'd go
our way undisturbed, he says,
gravity still there to spin the planets
like clockwork. Though not, of course,

any light. Life elsewhere? likely,
though it's our moon
that slowed us down enough for the local
show. And the missing matter
in the universe? Probably stashed
in all those black holes.
We file out soberly, thinking astronomers
take rather too long a view.

Control parameter

A dynamical system has variables that spontaneously change in time, and parameters that are often under your control but are held fixed and determine the nature of the dynamics. You can think of these parameters as control knobs that you can adjust to achieve the desired temporal behavior. They are like the knobs on a radio or television set that allow you to change the program to which you are listening. It is these parameters that are changed to produce the different artistic images of each type that appear in this book. Chaotic systems can be extremely sensitive to small changes in the parameters (as opposed to the initial conditions), and the chaos can disappear entirely if they are changed too much. In fact, the problem of finding parameters that produce chaos in a dynamical system is a difficult and unsolved one except in special cases.

THE PILLARS OF CREATION

They tower six trillion miles high in space,
seven thousand light years away — flung force,
nursery of stars, fantastic clouds of dust:

stuff of hydrogen and helium
heated by starshine,
surface boiled away to cold core

to accelerate the rest
into horseheads, elephant trunks,
the Eagle Nebula —

Pillars of Creation, we name an instant
of boiling heat and roiling cloud,
a bodying forth of the dynamic universe —

as each of us.

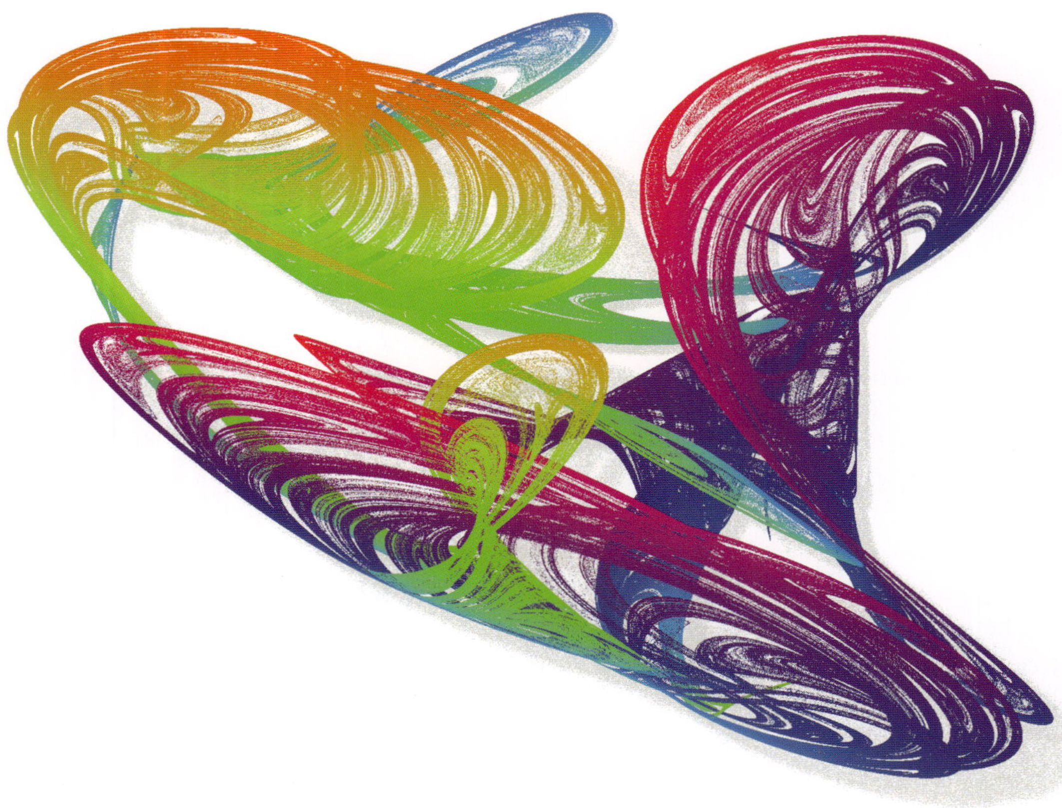

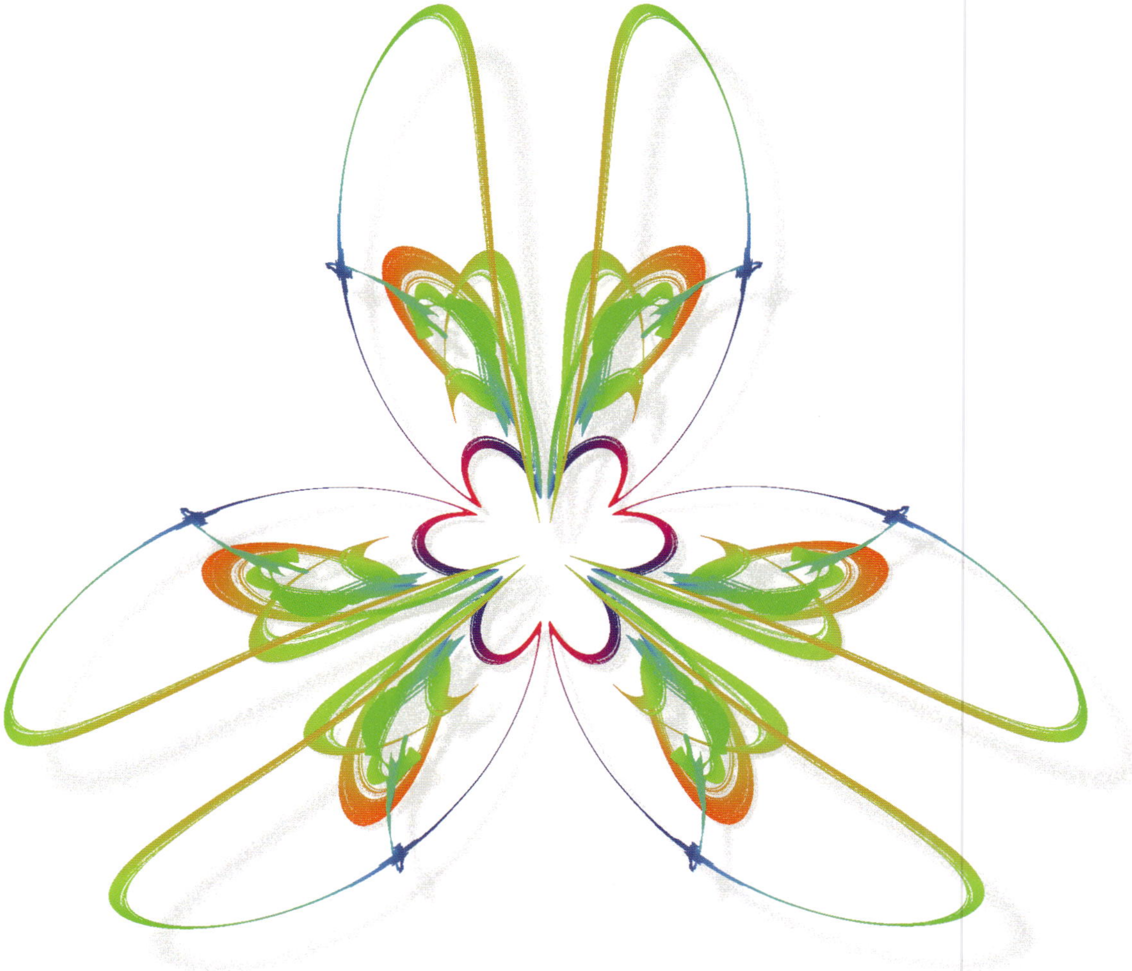

Chapter 2. Viewing Dynamics

There is no limit to how complicated things can get, on account of one thing always leading to another. — E. B. White

Where Do We Go Next? (Time Series)

WALKING TO WORK

You have time
  between one step
   and the next

To notice the sparrows
  feeding on the white berries
   of the red osier dogwood,

To discover the dew
  of the cold night still strung
   in droplets along the spiderwebs,

To see how the maple leaves
  shadow one another,
   kaleidoscoping in wind,

To admire the light-blue wild asters
  filling the neighbor's yard
   with slow-moving bees —

And you as slow
  in the brisk fall air,
   still a mile to go!

Time series

Dynamical systems are usually studied by making a series of observations of one or more of their variables in a sequence of equally spaced time steps. A remarkable fact is that such a time series of even a single variable can in principle reveal the complete structure of the dynamics, including whether the system is chaotic, how many variables are needed to describe the system (its dimension), and how far into the future predictions can be reliably made (the Lyapunov exponent). Examples of time series include daily stock market averages, meteorological data, electrocardiograms, and electroencephalograms. Detecting and measuring the determinism in such systems, even if they are chaotic, clearly is of considerable importance since that might allow a better understanding and more accurate prediction of the system.

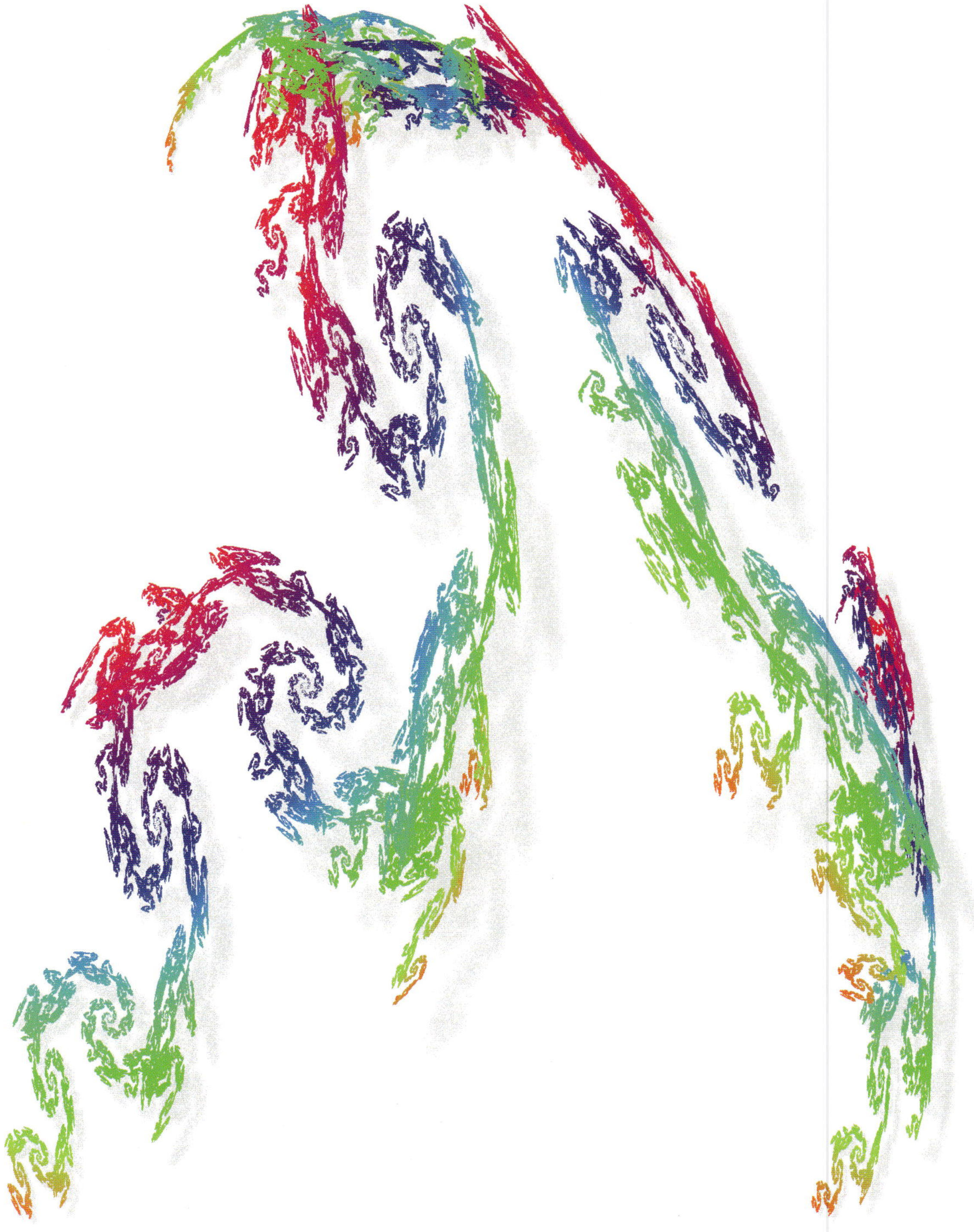

THE POETS READ FOR THREE HOURS IN THE GRAPE ARBOR

Light through the grape leaves
Could be any century's, swelling
The small green clusters
High overhead, waxy chandeliers
Suspended in the light and shadow
Of our living roof
On an afternoon
Suggesting rain again as the roses bloom;
Voices mix with the poems —
Family talk, babies' wails, traffic hum,
Bird song, the obliterating drone of planes
Making their last approach,
And the wind lulls
Some of the audience to sleep, or deeper
Listening — the vines are moving, now,
Touching the chairs; tendrils
Brush a shoulder,
Beginning their slow wrap up.

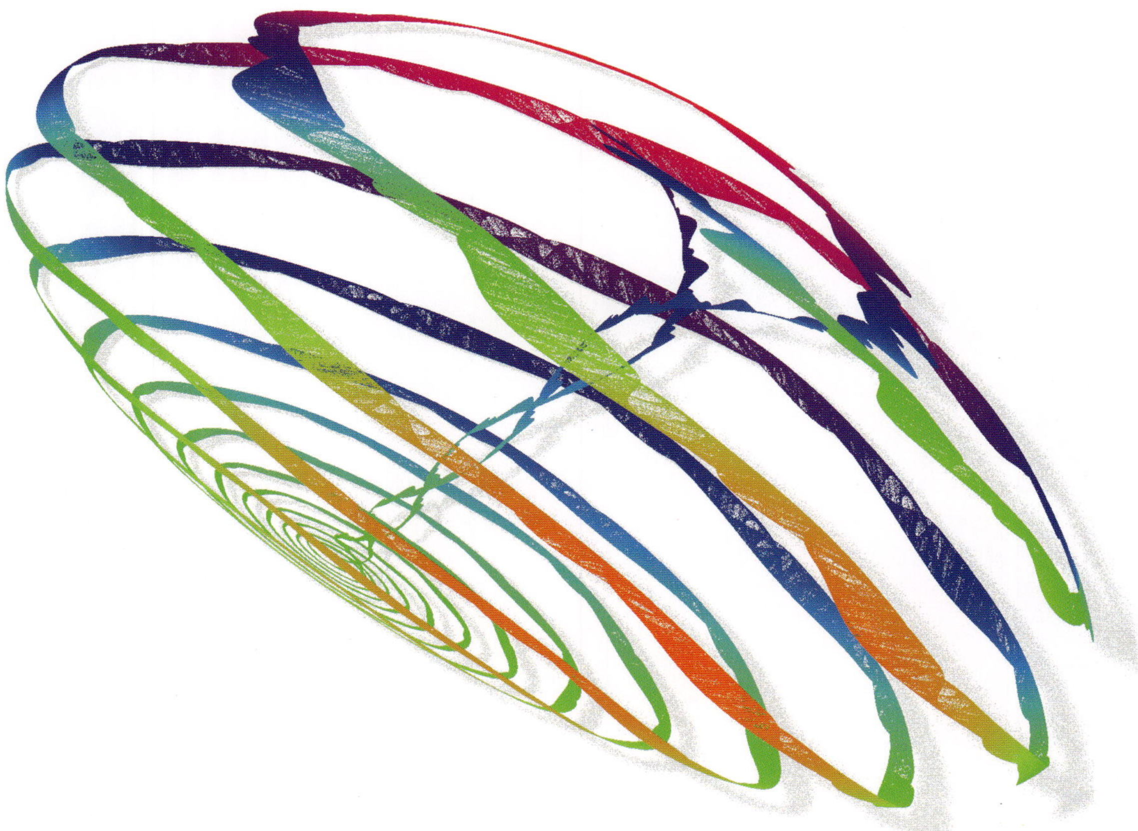

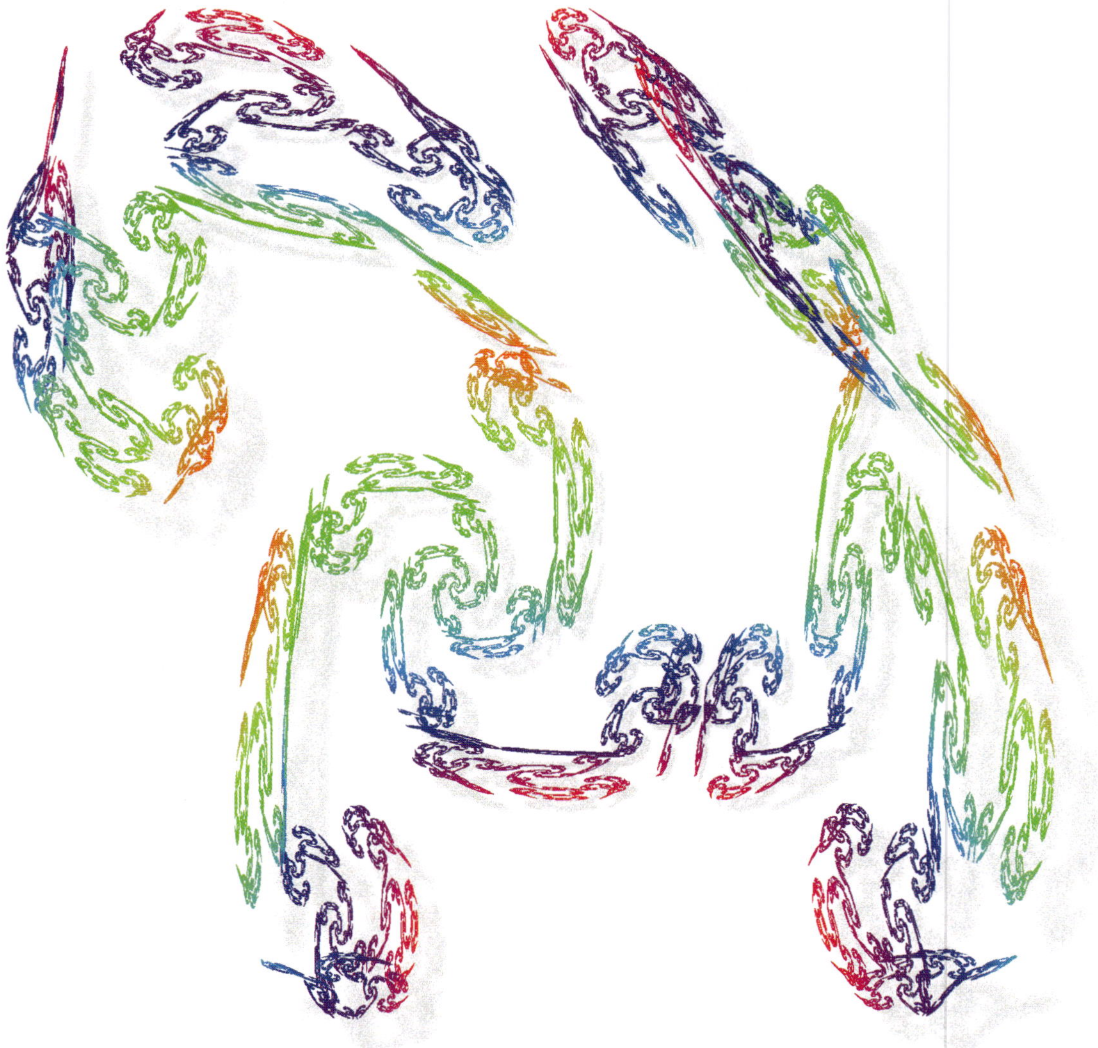

THE DOG GETS LOOSE

February rain has uncovered
All the winter tracks,
Etched out icy prints to reveal
Bare patches, sodden ruts,
And the day warm enough
To delight my hound's nose —
Criss-crossing the rich clues
He's working up
All the stories of winter —
Three days will hardly be enough.

LYING IN BED FLIPPING THE REMOTE CONTROL THROUGH THIRTY CHANNELS

I have wasted my life —
The Learning Channel is asking
whether the conehead skulls of Nazca
could be aliens from another planet
and grinding up bones for DNA to find out,
and the reality show's Joe Millionaire,
in reality a poor bobcat driver, is accused
on the talk show of being gay, or an actor,
and the alien engineer of a StarTrek clone
has a disfigurement that, on the channel
next door, could be cured while we watch,
live surgery in living color
for reconstruction of her eye socket,
and detective Colombo, yet again,
is remembering just one more thing
to ask the guy whom we've suspected
since the opening scene —
and the commercial mix of dogfood,
ab machines, and spaghetti strainers
somehow beef up and sluice the composite
as I sample 15 seconds each channel,
raising an electrostatic fog as thick
as any Bermuda Triangle.

Soliton

Most waves in nature, such as sound waves, light waves, and waves on the water, are linear because their amplitude is small. Linear waves maintain their shape and freely pass through one another without distortion, making it possible for dozens of radio signals to coexist or for conversation to occur in a noisy room. A water wave becomes nonlinear when its height becomes comparable to the depth of the water, such as when it breaks on a beach. Nonlinear waves generally perturb one another and distort because different parts of the wave move at different speeds. Solitons are nonlinear waves that move without distortion and freely pass through one another except for a slight time delay. Solitons were first observed in 1834 by John Scott Russell (1808–1882) who followed one on horseback along the bank of the Union Canal near Edinburgh for a distance of over a mile. They have subsequently been found in the solution of a variety of nonlinear partial differential equations.

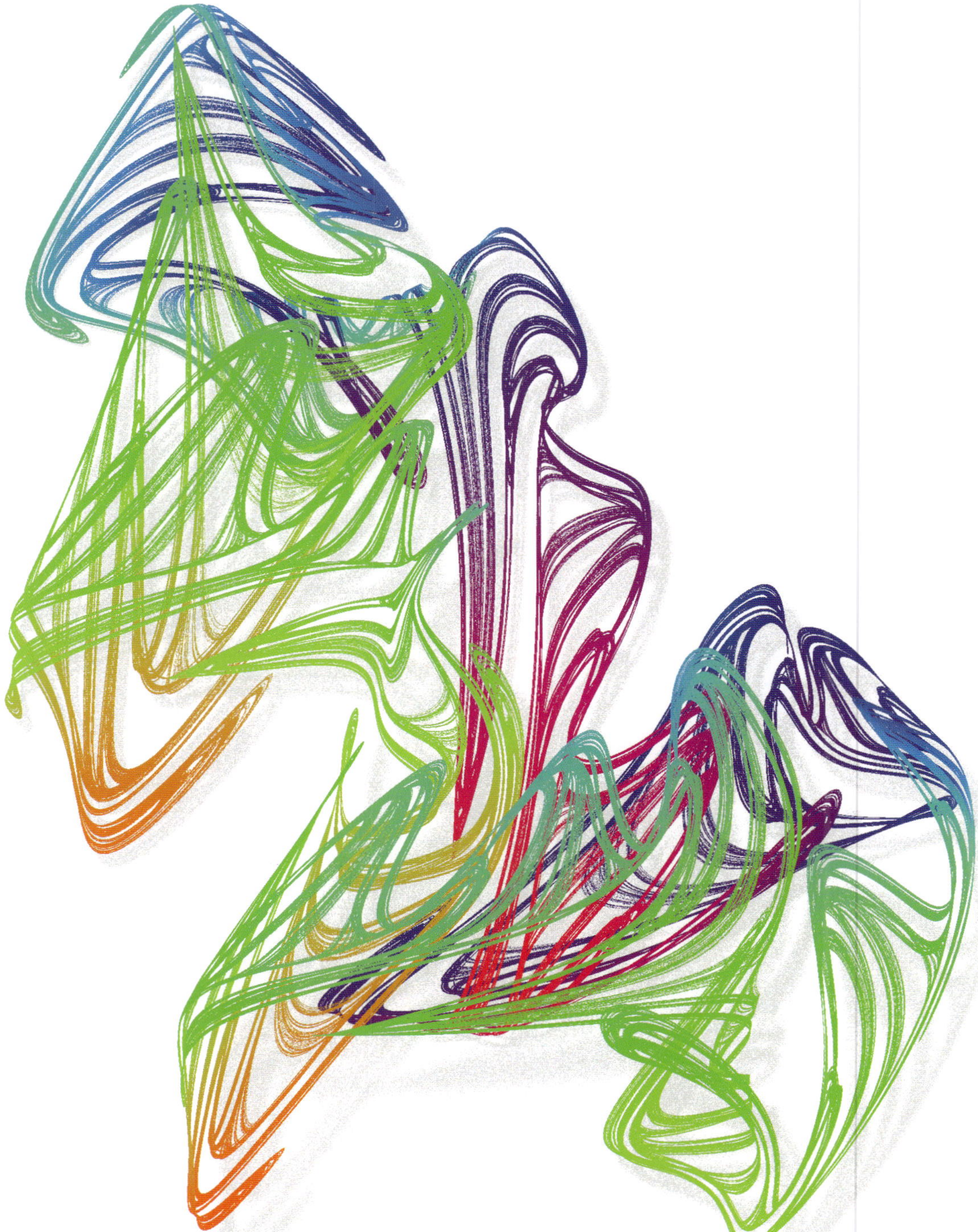

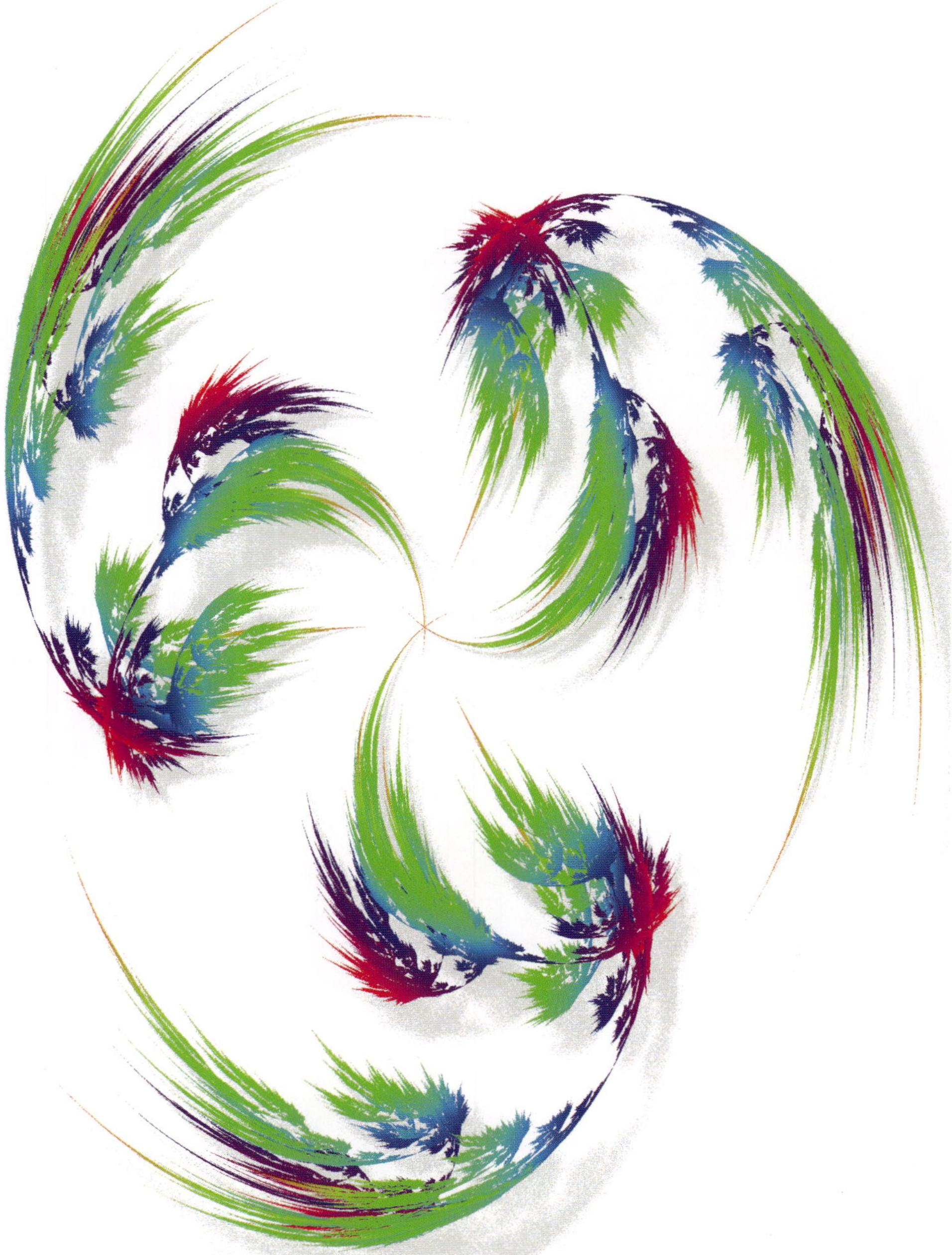

SUMMER CONTRA

Dancing, we learn the meaning of sweat —
not the polite shine of forehead
or underarm line of faint
perspiration but sweat, the full
outpouring drench of body as pump
as we whirl, gloriously cool, through
ninety humid degrees, sweat splashing
off hands that we slap in long lines
forward and back, sweat running in rivulets
down partners' necks, dripping from eyebrows,
noses, and chins as we swing in eyelocked embrace,
sweat soaking the shirtbacks, cool wet
to the touch, as we cast off our neighbors
for ladies' chain, offering our slippery
grips, hair plastered to forehead and nape
as though caught in a sudden summer storm,
the fiddlers cranking up the pace
as we weave and drip through the hay
for four, circle left, swing once more,
thank the band, rush for water and another
partner — walk out into heat, two hours later,
washed salty and clean from every pore.

SUCCESSION

Now popples blaze as yellow
As the goldenrod, as the late
Sunflowers; the sumac
Is considering
Whether its time has quite come;
Reeds take on the sober brown
Of mother mallards; we hardly notice
How green is leaving, has left, is gone.

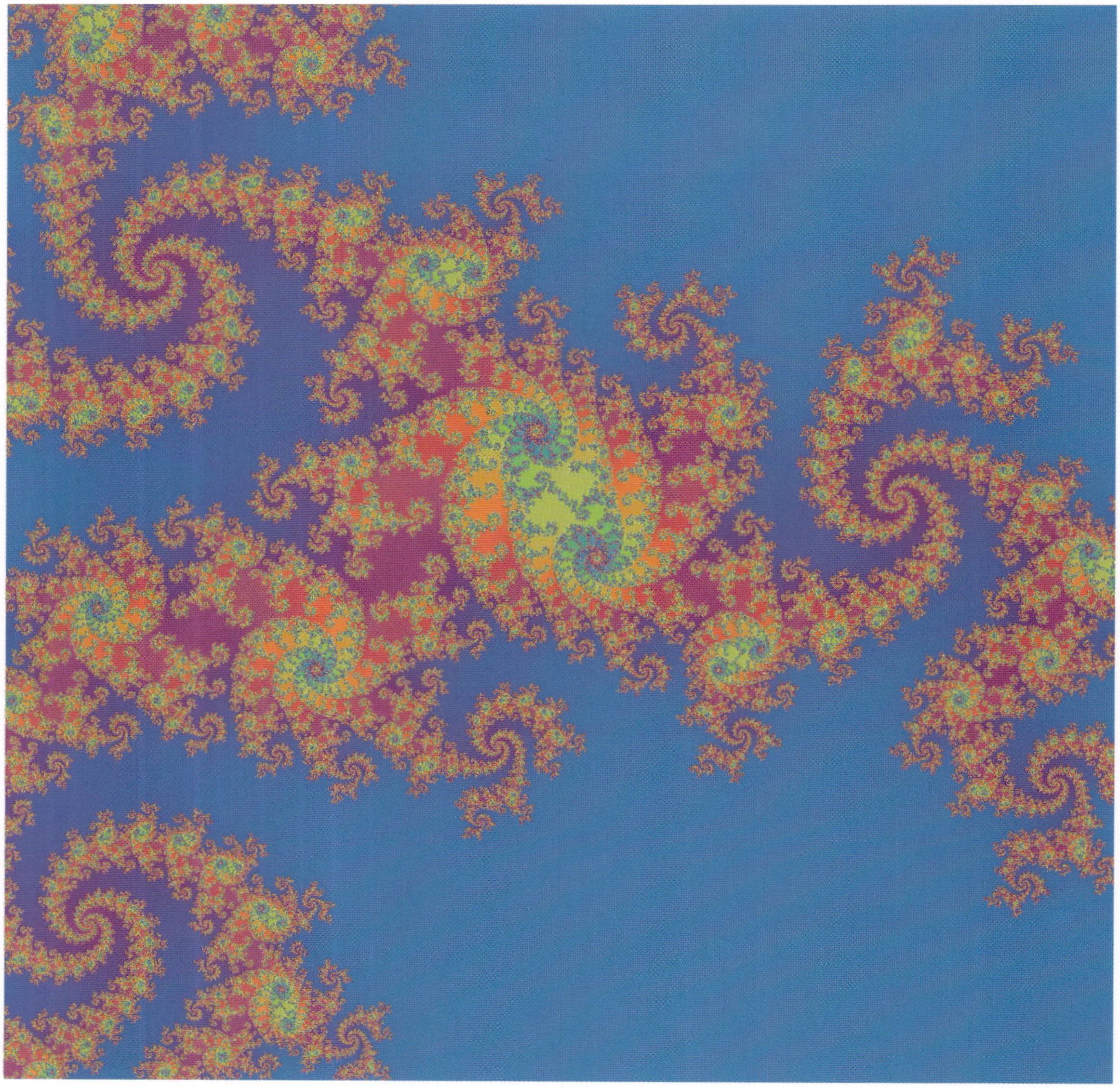

SUDDENLY THE SUMAC

Among the white asters the monarch feeds,
Her closed orange wings as brilliant
In sun as the sumac, suddenly red;
Till, letting go, she is caught up
In the wind; and though she soundlessly
Turns and turns, back toward
The burning white branch, she is no match
For the larger shadow of her desire,
The river of air moving south.

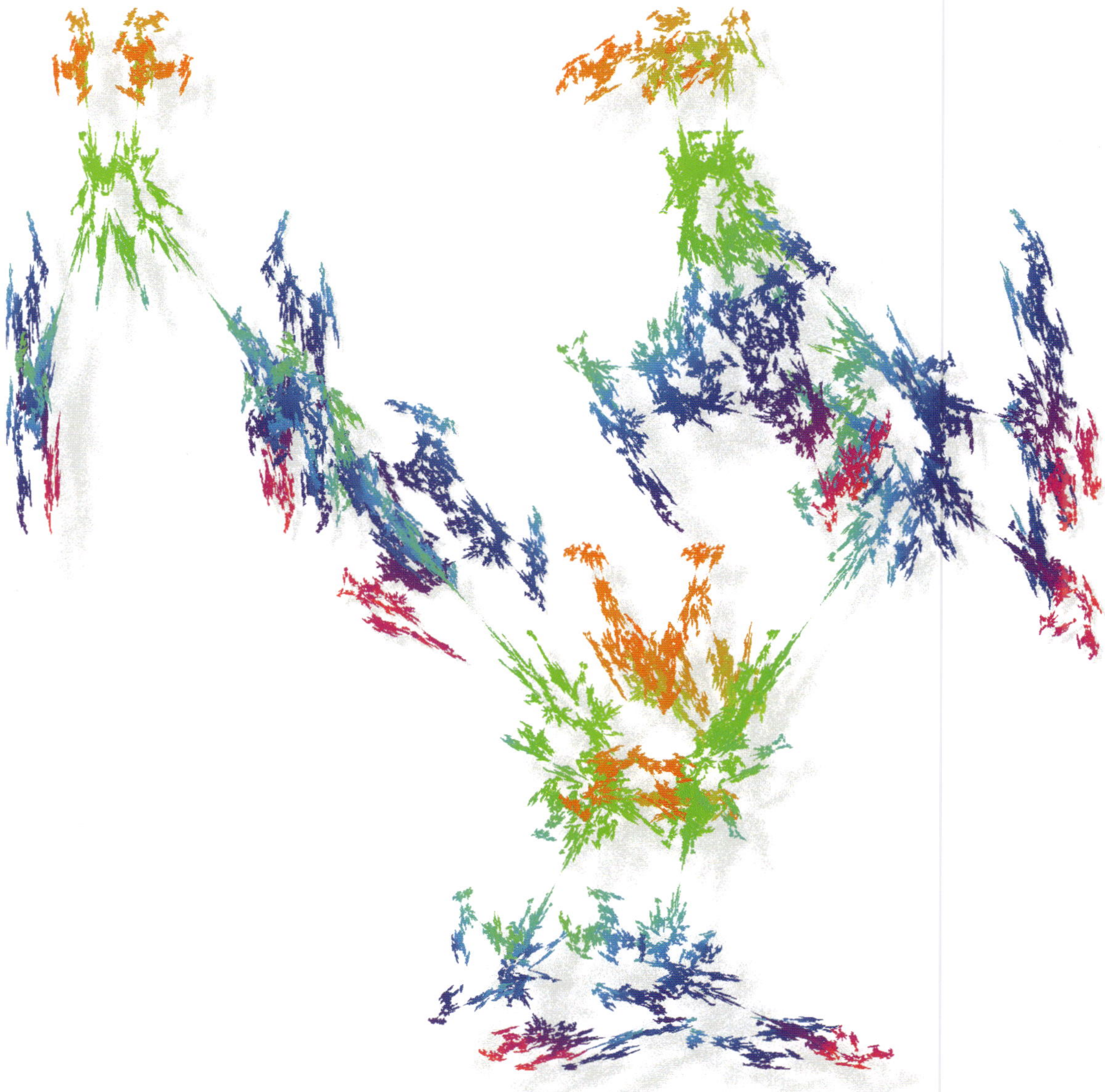

NOVEMBER CANOEING

Early, and Will and I ease the canoe
    into the cooling water
        of Lake Wingra, lightly,

the way a sable brush moves forward
    with no purpose but its own motion
        trailing color

as we allow ourselves
    to push off from shore,
        follow the ins and outs of currents,

thread our way past the springs,
    the great heron who stands sideways
        on the brushy shore,

neck folded, motionless.
    The mallards scuttle before us,
        showing the only way out of the slough.

Back along shore, among grey–black–beige
    of bark and grass, rush and thicket,
        what we remember is the hunched

blue shoulder, the white eye streak,
    the sharp yellow bill poised
        above watery blue shadow.

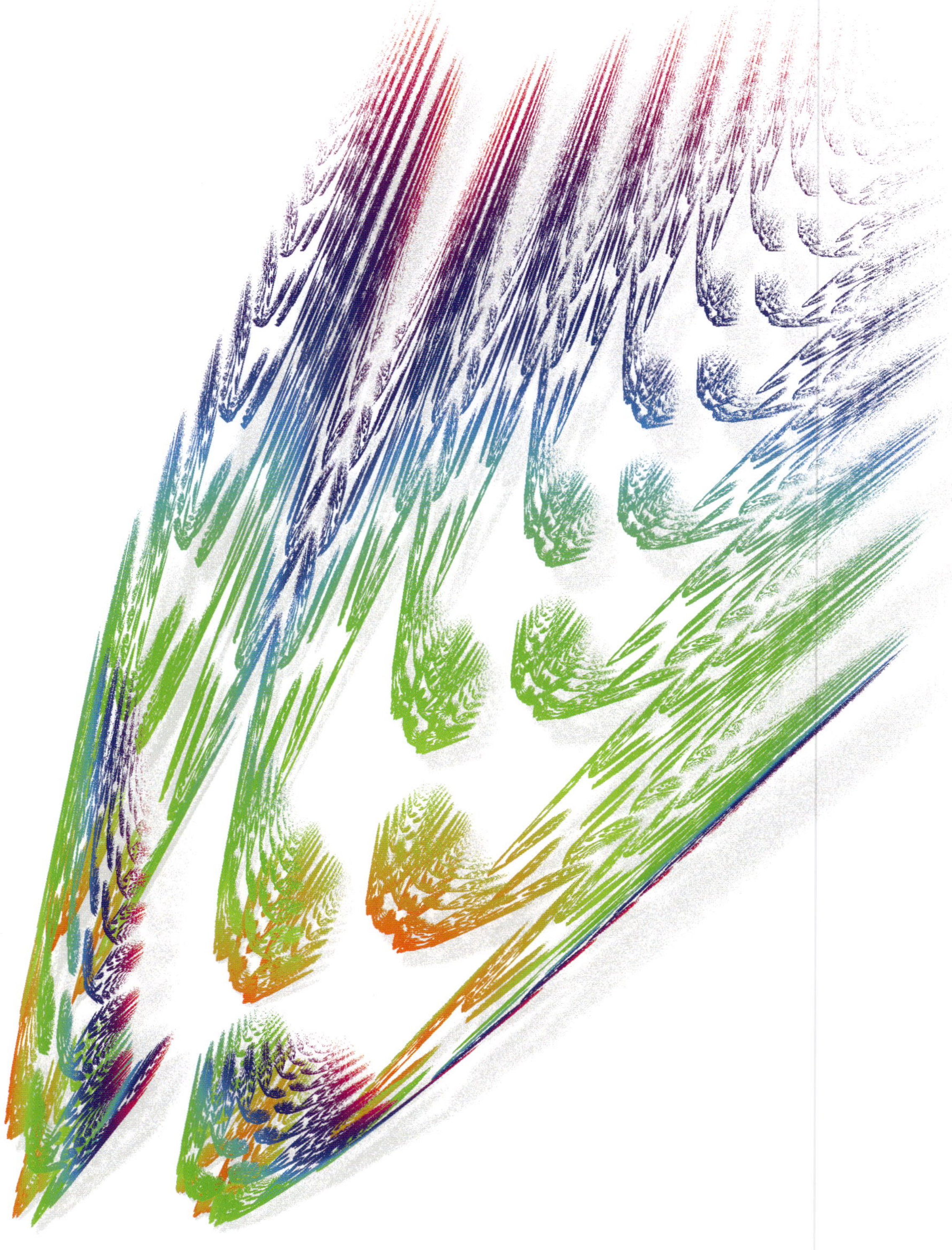

GOING FORTH

I take the mountain path.
Cinders steady winter boots.
Sun makes white gold
Of the mountain tops.

Luxuriant, this steadfast
Loneliness, this true ease,
Feet over and over stitching
Blue hollows in white crust.

It means seeking,
This going forth. On black
Wings the ravens search
Through the sky's blue scarf.

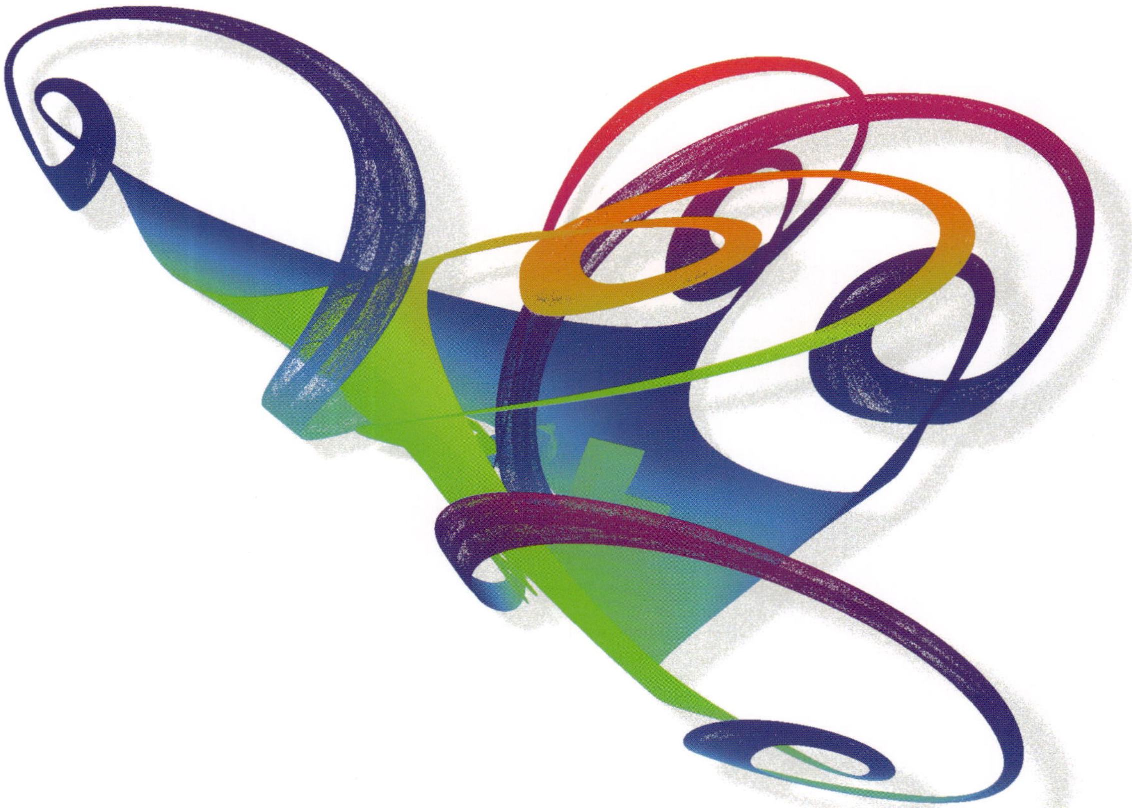

Have We Been Here Before? (State Space)

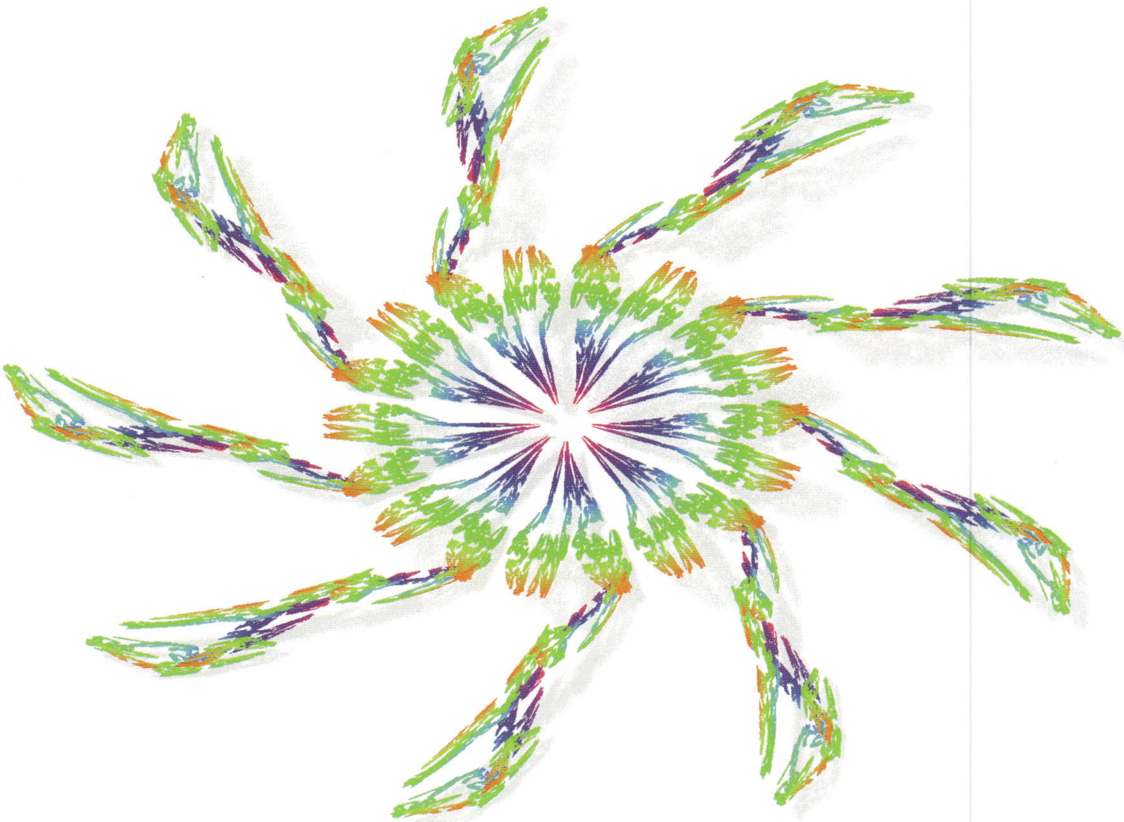

State space

State space is a generalization of the real three-dimensional space in which we live. A dynamical system has a number of variables that are simultaneously changing in time. We imagine these variables to be different perpendicular directions in state space, and the behavior of the system is then described by a trajectory in this state space, much like a fly buzzing around the room in real space. State space is not limited to three dimensions, and in fact it can be infinite-dimensional. Space with dimension higher than three is also called "hyperspace." When the state-space variables are the positions and velocities of moving objects such as the molecules in a gas, we use the term "phase space" rather than state space, although these terms are often used somewhat interchangeably.

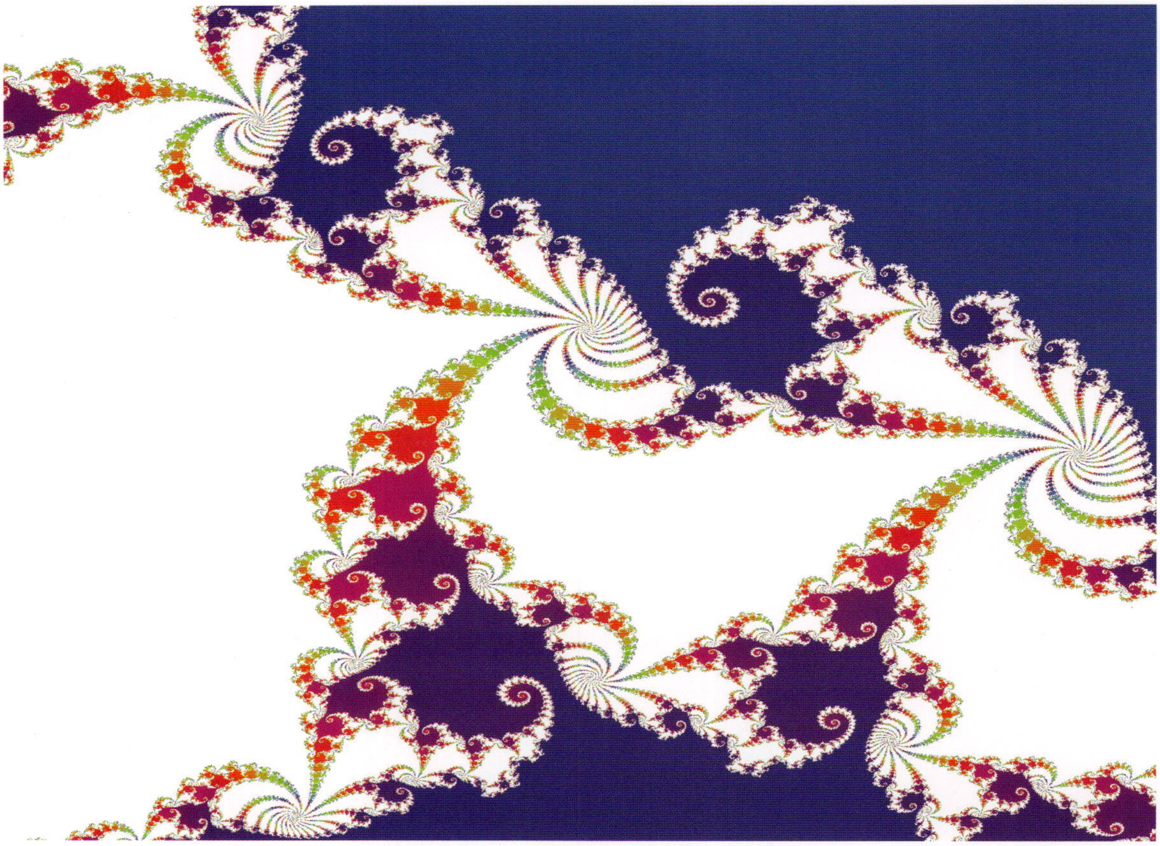

NEW YEAR'S EVE
BANFF, THE WOODS

Dear Ones — I make no resolution lists,
listen instead to Chopin, Holiday, take
lunch with friends deciding yes, the squid,
salmon pate, cold asparagus, roast beef,
horseradish sauce, yes, bread pudding please,
cherries flambe, and yes, more orange juice,
more black coffee, pass the cream —
we give ourselves to meat, to drink,
to a little of everything. Back in the day,
we hike Tunnel Mountain to the peak,
see Mt. Rundle flare up in the fireworks
of the sinking sun. The Spray River
runs on into the Park's true wilderness where,
as here, the ravens watch what goes, what comes:
the old year. The new one. Climbing down
through blue dusk, the whirr of grouse,
we make our way by icy switchbacks.

CHICKADEES AT ANVIL LAKE

Their scratchy feet curl into our outstretched palms,
curve like newborn's fingers around our own
as they take the black oil seeds from our hands
and fly away to birch and arborvitae
to eat from lifted claws the sun's packed energy.
Trees buzz and twitter with those who wait.
We there beside the smoky warming house
take off our skis and gloves to scoop the seeds
and stand, arms out, in snow and cold,
a Francis of Assisi flock awaiting birds.

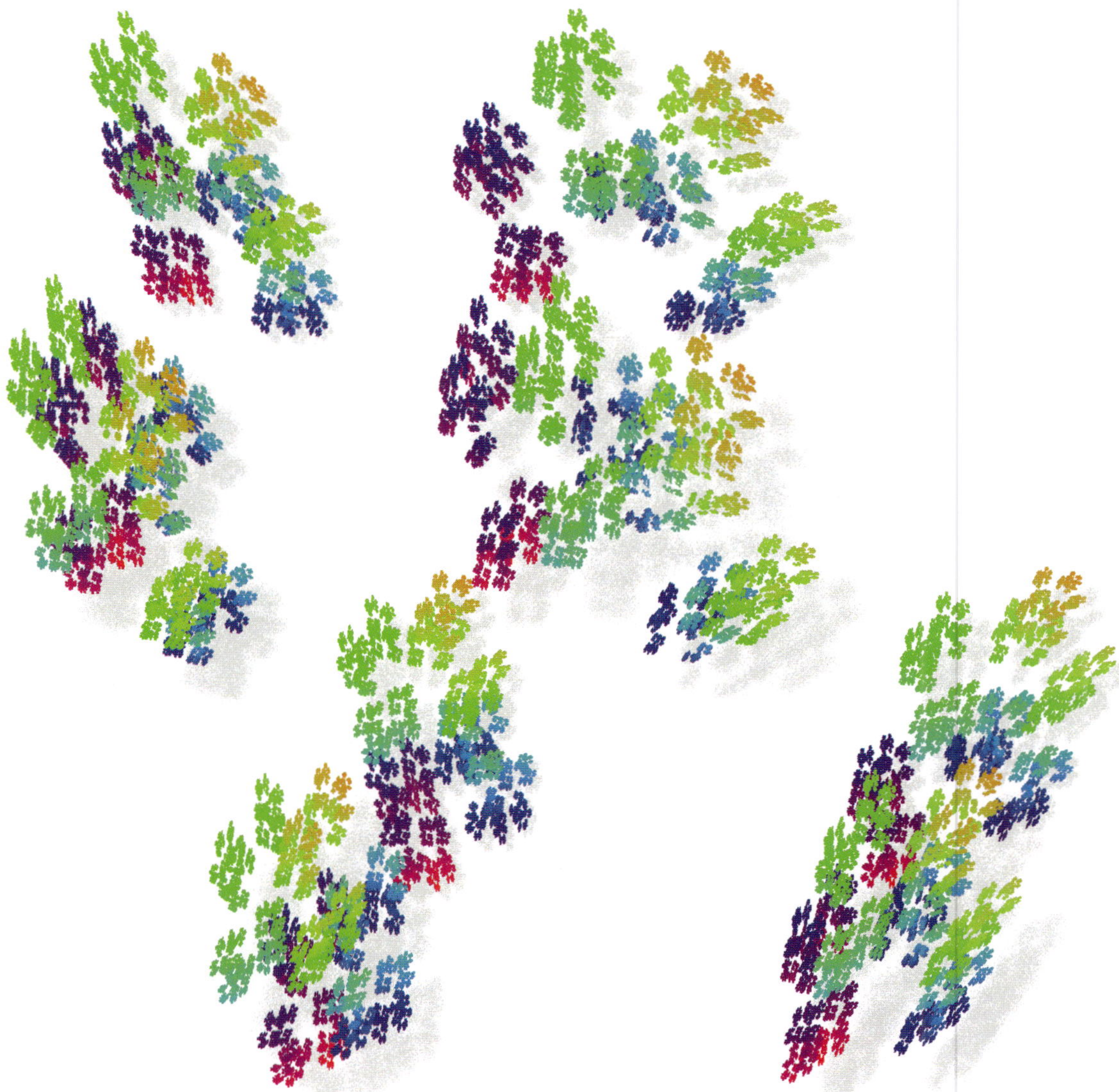

SPRING BREAK-UP

We go in search of the wind's work,
its howl shifting us free of winter's grip —
wet leaves skirl down the streets,
branches that held on through ice
and snow's weight break and scatter
in its blast: south wind, gale-force,
bearing the warblers and kinglets north.

Sandhill cranes hunker down in the marsh.
The few ducks taking flight
fight to get back to earth,
and Lake Mendota, iced over this morning,
runs with white caps now
as we circle southwest shores to come
to the northeast edge where clinking chimes

rise above the roar — ice, honeycombed,
three inches thick, driven to shore,
cracked in the wave rush, broken to bits,
piled in drifts above our heads — *ice shove,*
it's called — we climb and slip its avalanching
pitch, this side a playground, that side
hypothermic death, as sun haloes the ridge.

Back home, crocuses punch saffron and purple
through the strangle of leaf litter.
Bees already hunt among the orange anthers,
petals giving way to the wind's hot mouth.

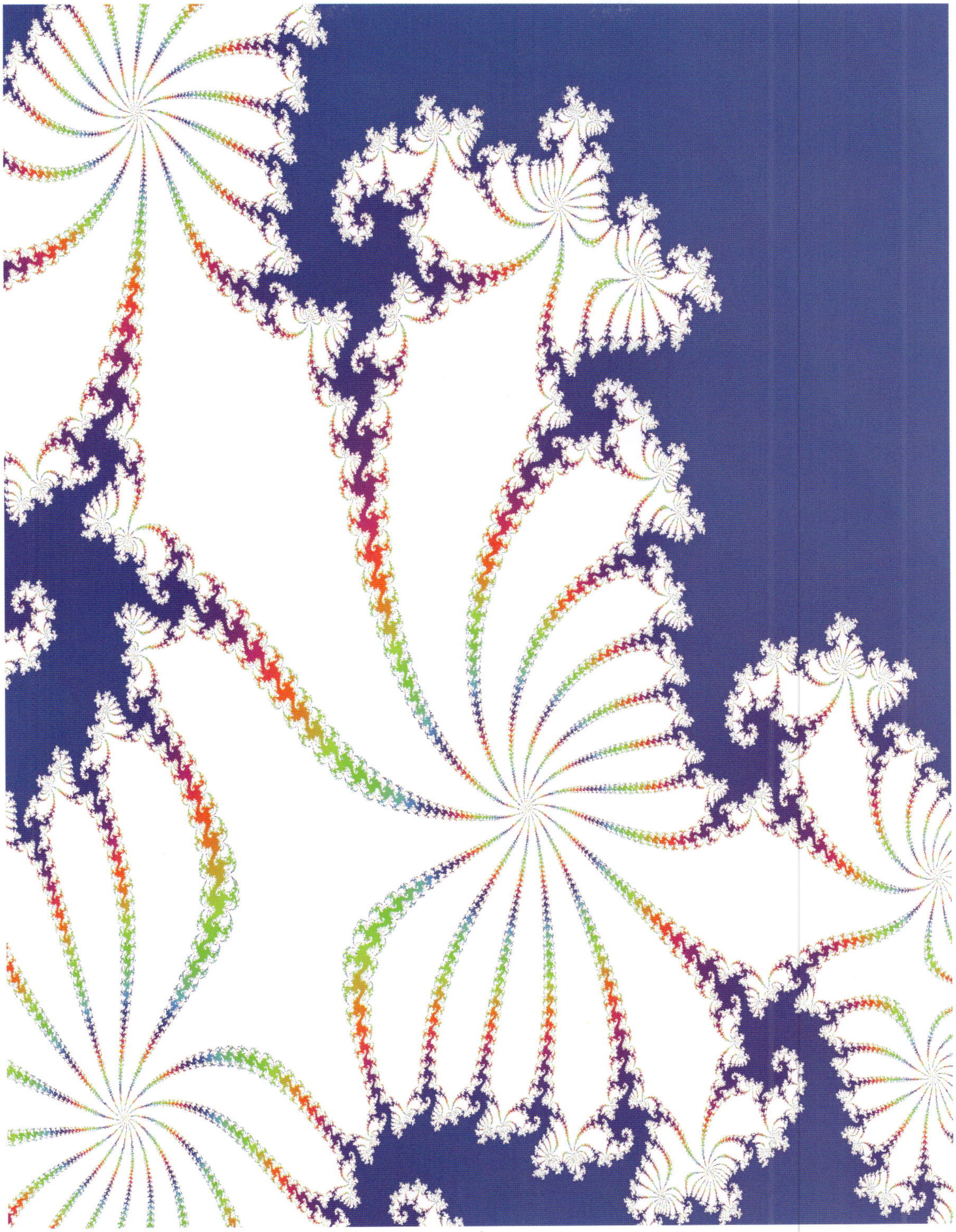

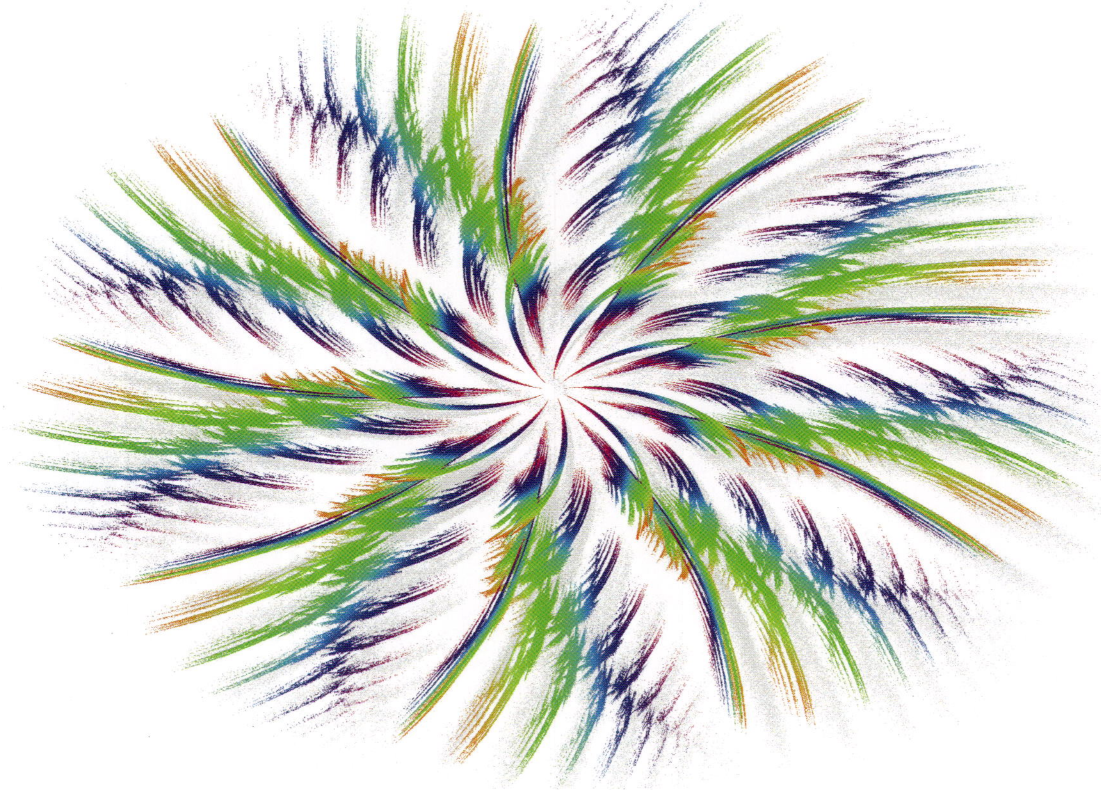

THE GATHERERS OF STORIES

for Harold Scheub

Every day I drive past
The professor who walks to work,
Who walked for years
Through Africa, gathering
Stories from the tellers
In dusty villages — I know
The stories repeat in his head
As he walks toward a midwestern
Lecture hall where hundreds
Of faces lean forward
At his opening gesture
As, one by one,
Voices begin to speak
Their stories through him,
The common rhythm of walking
Pacing every translated word —
See how the banyan tree
Has canopied the room.

NORTHERN PIKE

Water so shallow and clear
        I can see his unblinking eye
and yellow snout. Black dots
        on the sunfish minnows
schooling the water weeds.
        Their flicks of fear below
the wind-rumpled surface.

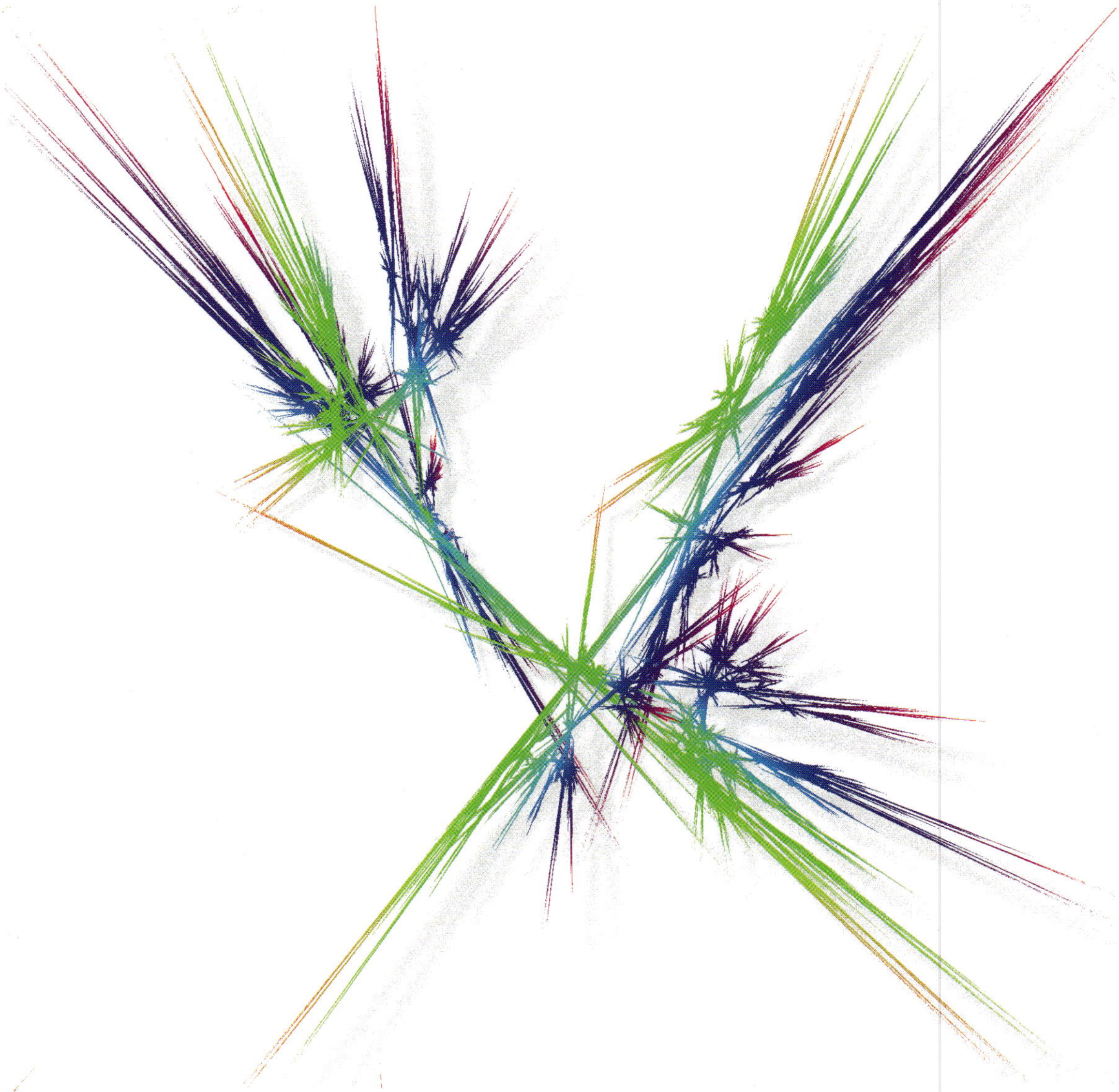

FIELDS

Walking the path by the field let go, beyond
  the shorn green where girls and boys,
    red or blue-vested, practice the moves

of a soccer game,
  tracking each other, the ball,
    all legs and eyes and cage of goal,

summer descends, green and pliant,
  at any moment of being, something flowering,
    growing, ripening —

in the grasses too deep to run in,
  where seedheads nod
    their varied nods, stiff or delicate,

yes or *no* to the wind
  that plays through them, and underneath,
    forking catchfly, cloverleaf,

I see two blonde heads, like flowers, visible,
  that first I take for lovers, and, closer,
    see are a boy, awkward and slow

for his height, and a girl, setting out food
  and water. She holds the cup, wipes
    his mouth. The boy, who has picked

a stem of grass, is putting the wrong end
  in his mouth. She turns it around,
    teaching the move of nonchalance,

the posture of picnickers in the grass.

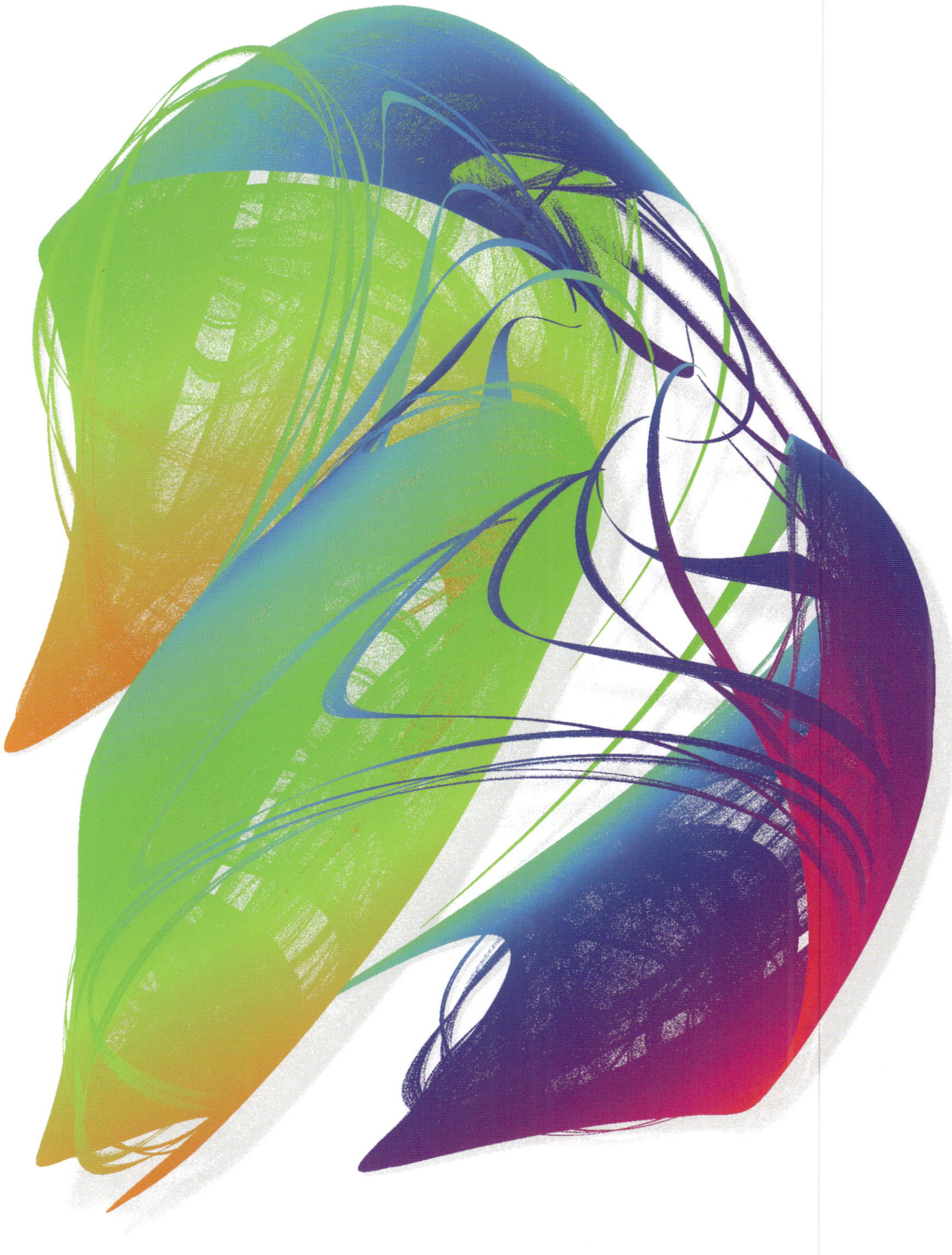

EOPSALTRIA AUSTRALIS: DAWN SINGER

Almost as modest and missable
From the back as her small,
Lichen- and moss-studded nest
Tucked in the green welter of forest,

She perches sideways to feed her fledglings,
Raw and gray, cupped deep in the nest,
Who open their tiny bills to show
The color of yolk rimming their throats

And see, arriving, rising above
Their horizon of sticks, the brilliant
Sunburst breast of their gray and dark-eyed
Parent — one day they, too, will open

Their throats in rainforest dark
To pour out the burst of notes
That sing up the first
Brilliant yellow of the sun.

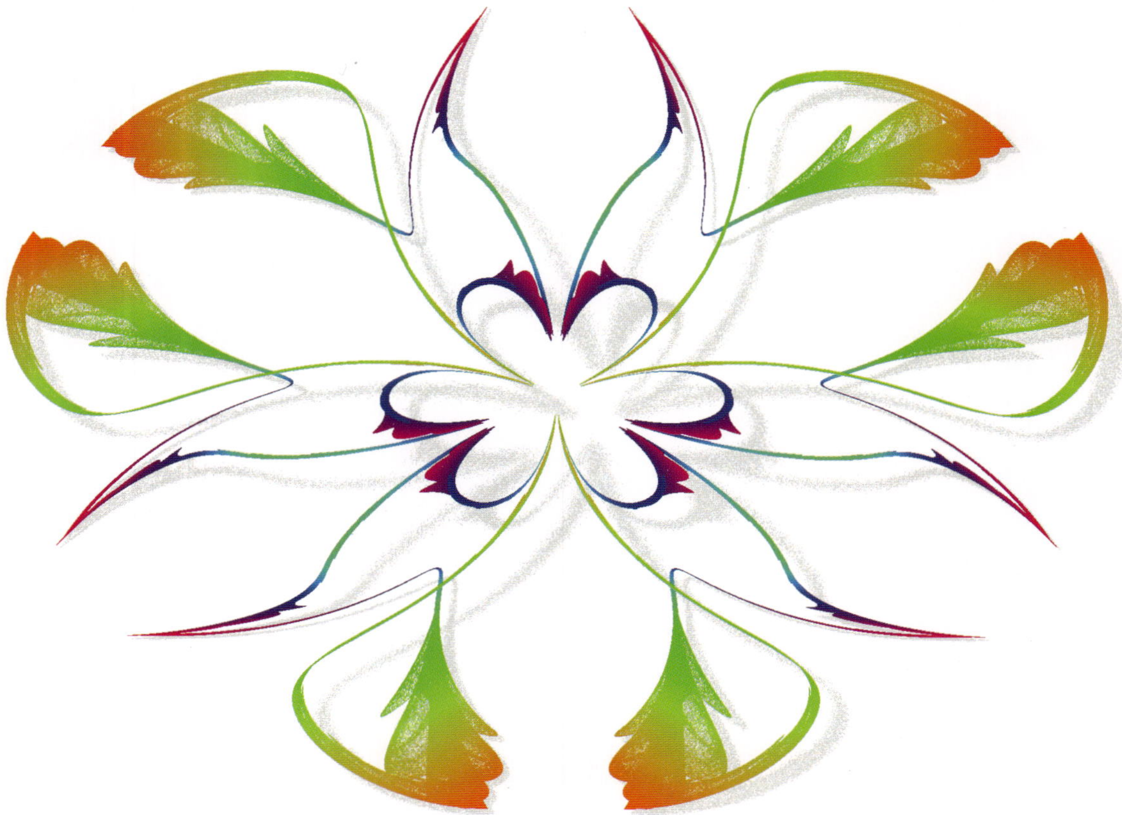

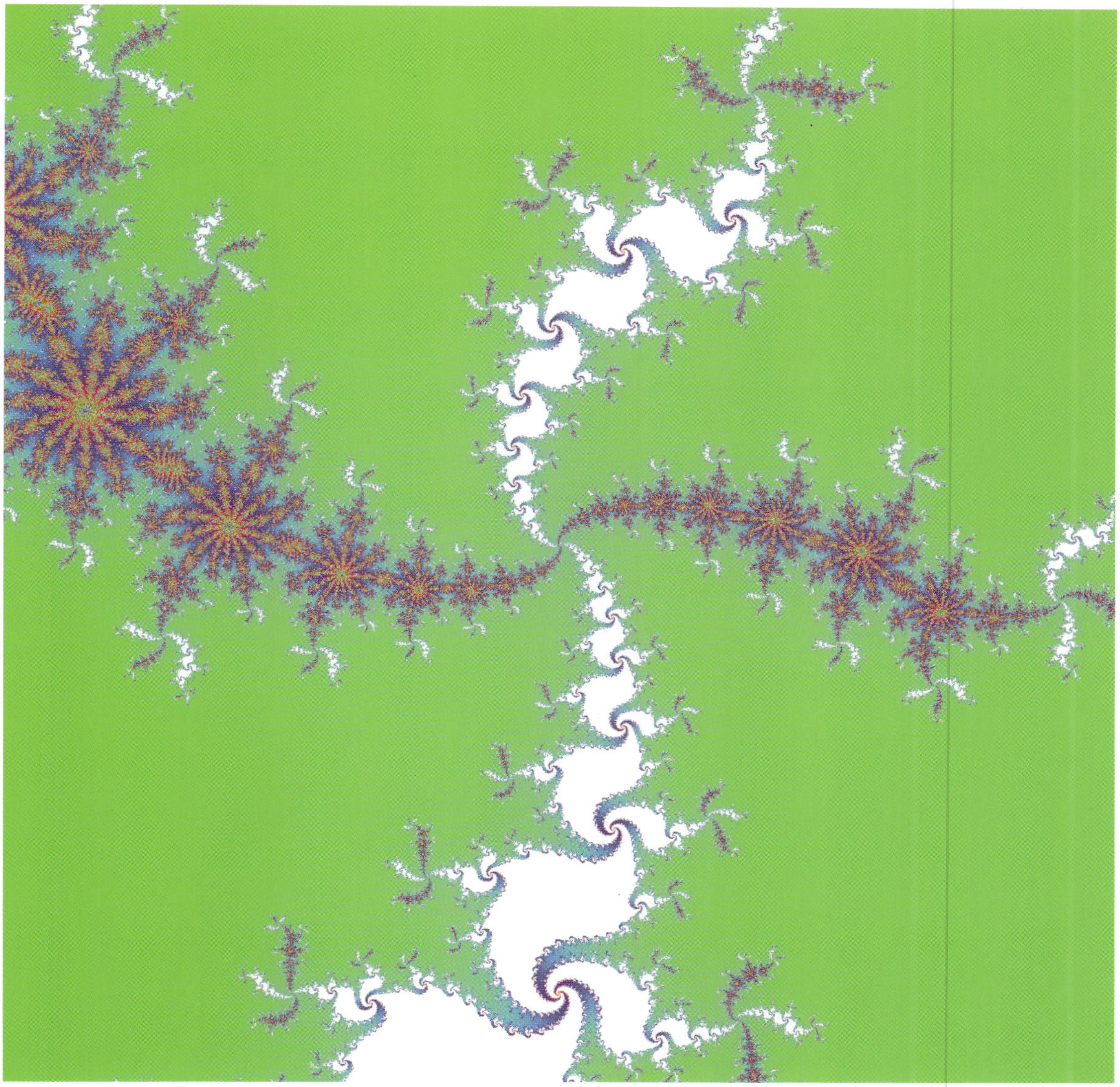

Chapter 3. Where It All Ends

If you don't know where you're going, any road will take you there.
—Doug Horton

Attractors

Attractor

An attractor is the set of points that represent possible final states of a dynamical system whose initial conditions are within its vicinity (its basin of attraction). An analogy is rain that falls within the watershed of a lake and eventually ends up in the lake, so that the lake is the attractor for the rain. Attractors can be isolated points (zero-dimension), closed loops (one-dimension), or tori (two-dimension), which are shapes like the surface of a doughnut or inner tube. Higher-dimensional tori also exist but are hard to visualize. Chaotic systems usually have strange attractors, which are fractal objects with non-integer dimension and structure on all size scales.

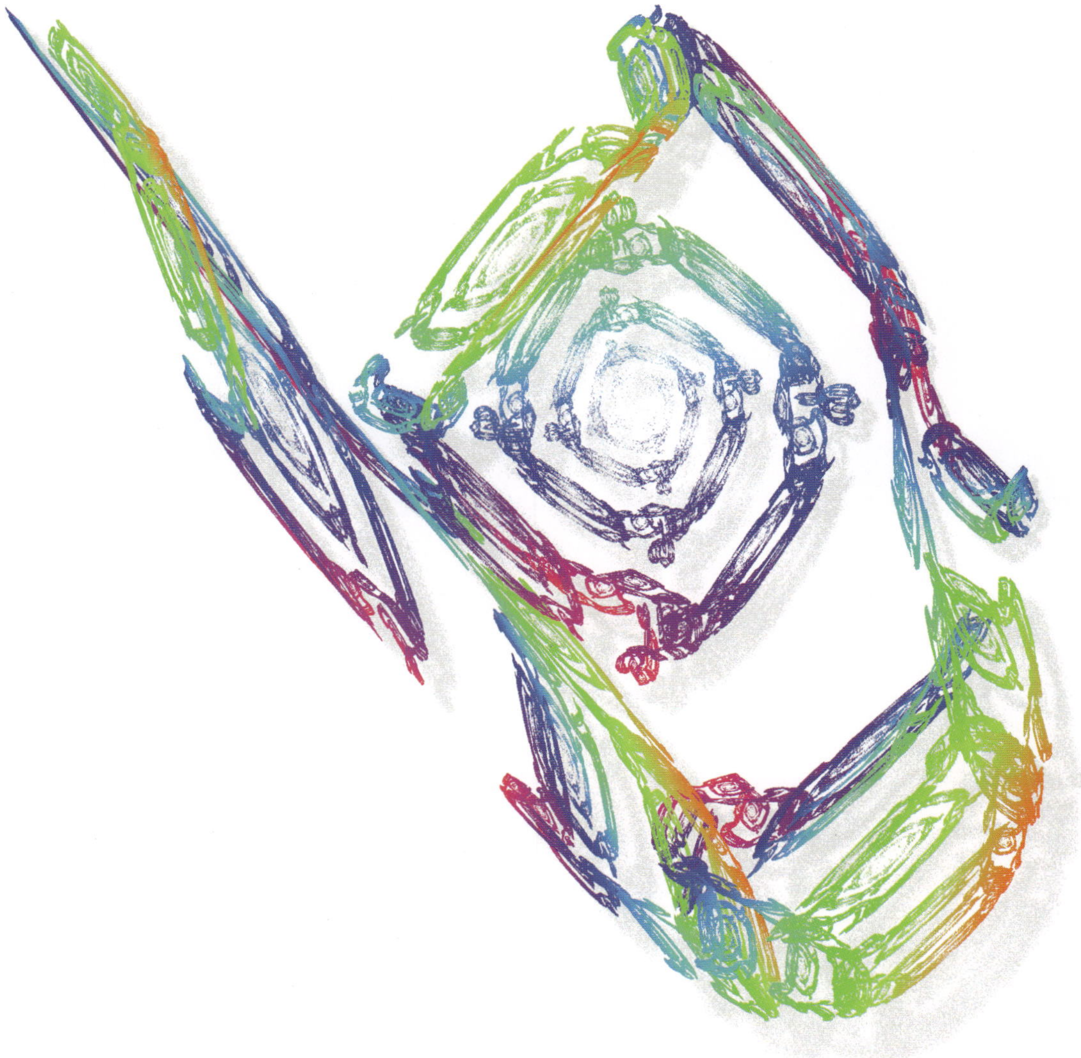

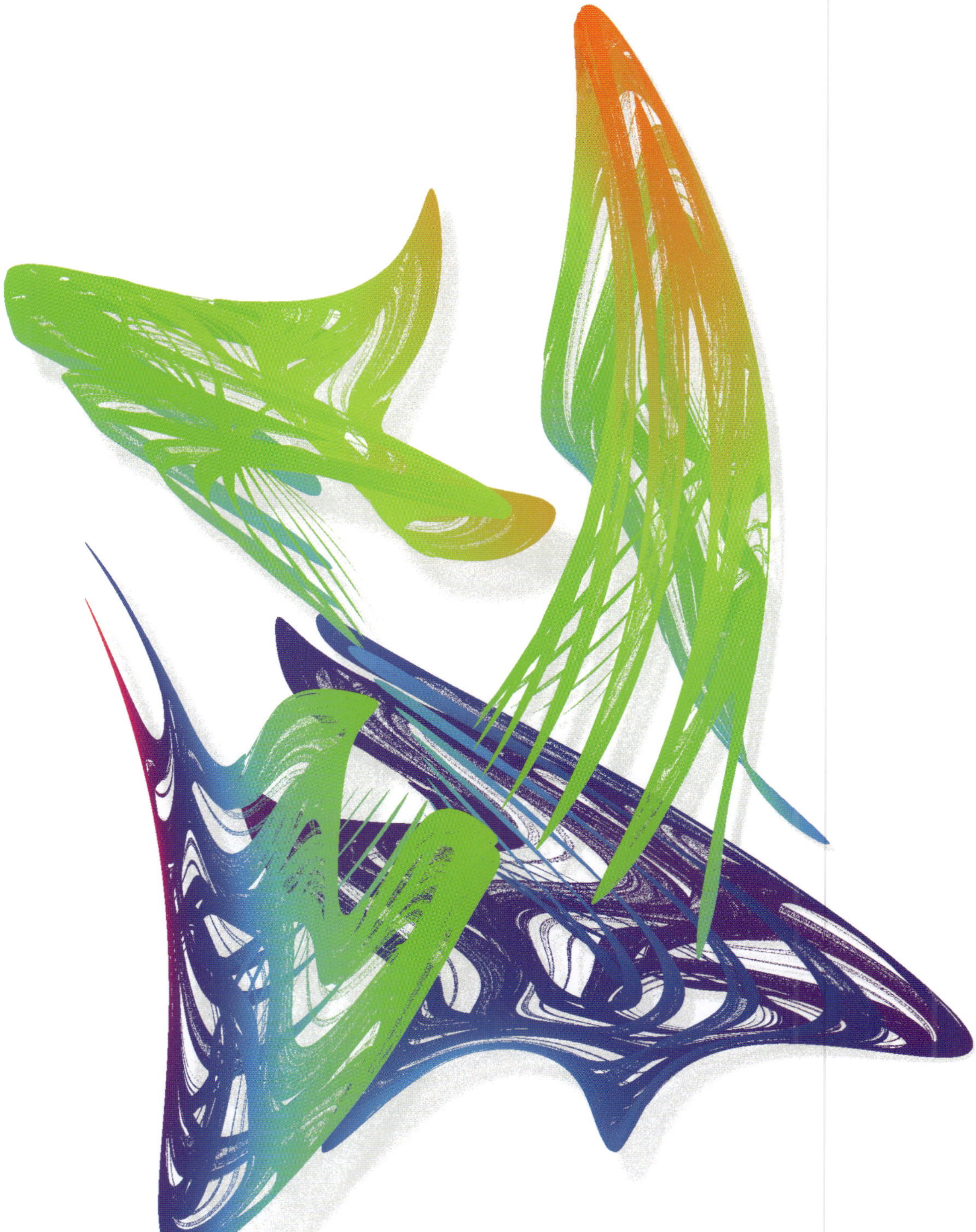

Stillness (Fixed Points)

THE FIXED POINT

The dot: how it stops
everything.
Finishes the thought.
Ends the sentence.

Where everything vanishes
in the end.
Period.

But it is not
the all of it,
though all come to it.

It is only the idea of no
dimension
over which we exclaim,

the vanishing point
that lends the observer
perspective,

a fiction
of the eye too far away to see —
speck, mote, egg.

Entropy

Entropy is a measure of disorder. There are many ways the molecules in a gas can be distributed, but the most likely way is for them to spread out into the most random and disordered way. They never spontaneously pile up in one corner. The second law of thermodynamics says that the total entropy of an isolated system always increases. This law is only a statement of probability, but the probability that entropy will spontaneously decrease is extremely small in a system containing many components such as the molecules in a gas. This argument has been used as evidence for intelligent design of the Universe, but there are many counterexamples. When a system self-organizes, such as the formation of an ordered crystal when water freezes, it is because the system is far from equilibrium and there is a considerable throughput of energy. Such a system is not isolated, and the total entropy of the Universe does increase even while decreasing in some local subset of it. Systems in equilibrium have maximum entropy because that is the state with highest probability.

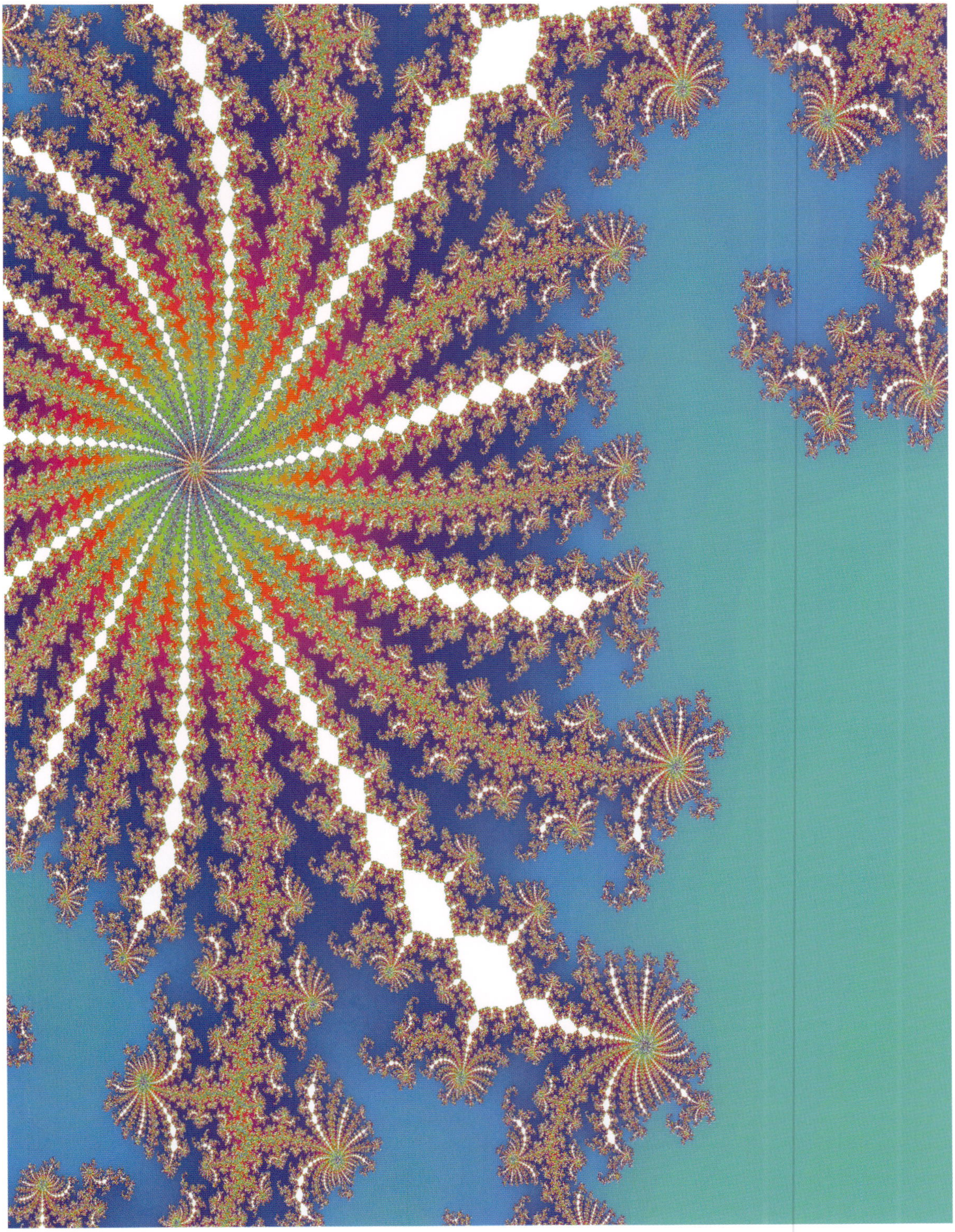

THE SCIENCE OF LIFE

in memoriam, Howard Temin,
discoverer of reverse transcriptase,
messenger RNA

What lasts? He taught us
That the codes for life itself

Could be rewritten and passed on —
What the cell's mutable core tells

The neighborhood
Whispered abroad by messenger

To liver and lung —
Those distant smoky cities

Where rumors grow.
When lungs reported their new story

He did not alter the days
Of his life, heart's delight, mind's eye.

Planted daffodils
To welcome the spring migration.

Equilibrium

A system is in equilibrium if the net force on it is zero. Equivalently, a small perturbation to a system at equilibrium does not change its energy significantly. The equilibrium is stable if the system tends to return to the equilibrium state when perturbed (think of a marble resting at the bottom of a bowl), and otherwise it is unstable. A system in equilibrium is often static, but there are exceptions such as a ball rolling across a flat surface, which is a state of dynamic equilibrium. A lake with a smooth and level surface is a stable equilibrium. Waves on its surface represent a perturbation, but they diminish when the cause of the perturbation is removed. Although equilibrium is the natural state to which an undisturbed system evolves, many important systems in nature are far from equilibrium, typically because there is energy flowing through them. For example, sunlight is the source of energy that allows a rich ecology whose stable equilibrium corresponds to the death of all species, and food is the source of energy that allows humans to organize and perform complex tasks.

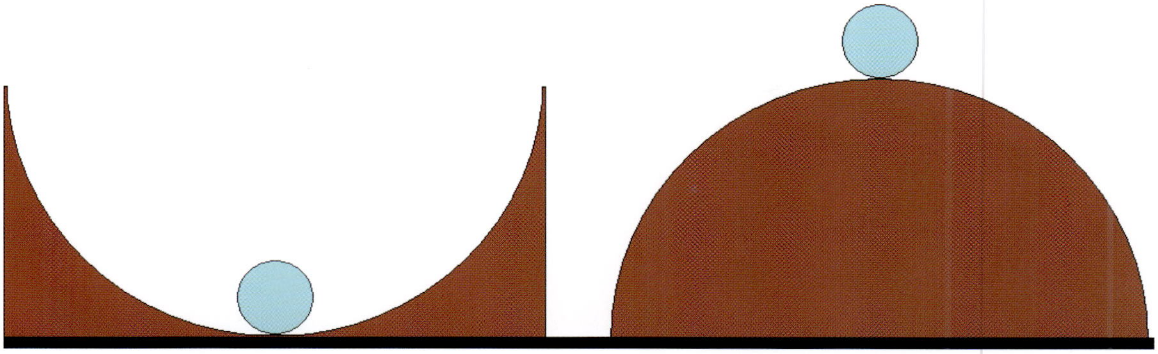

Stable and unstable equilibria

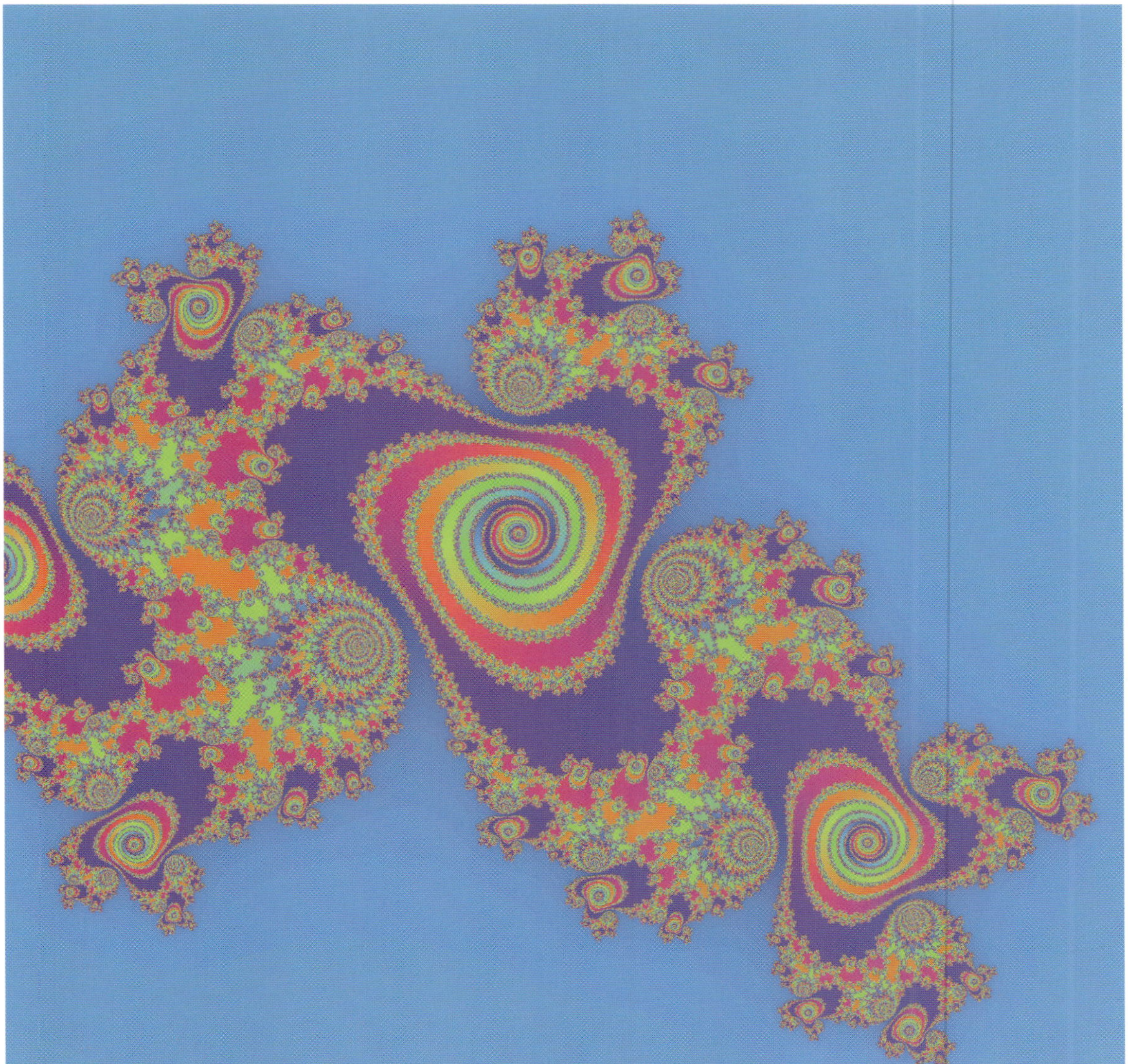

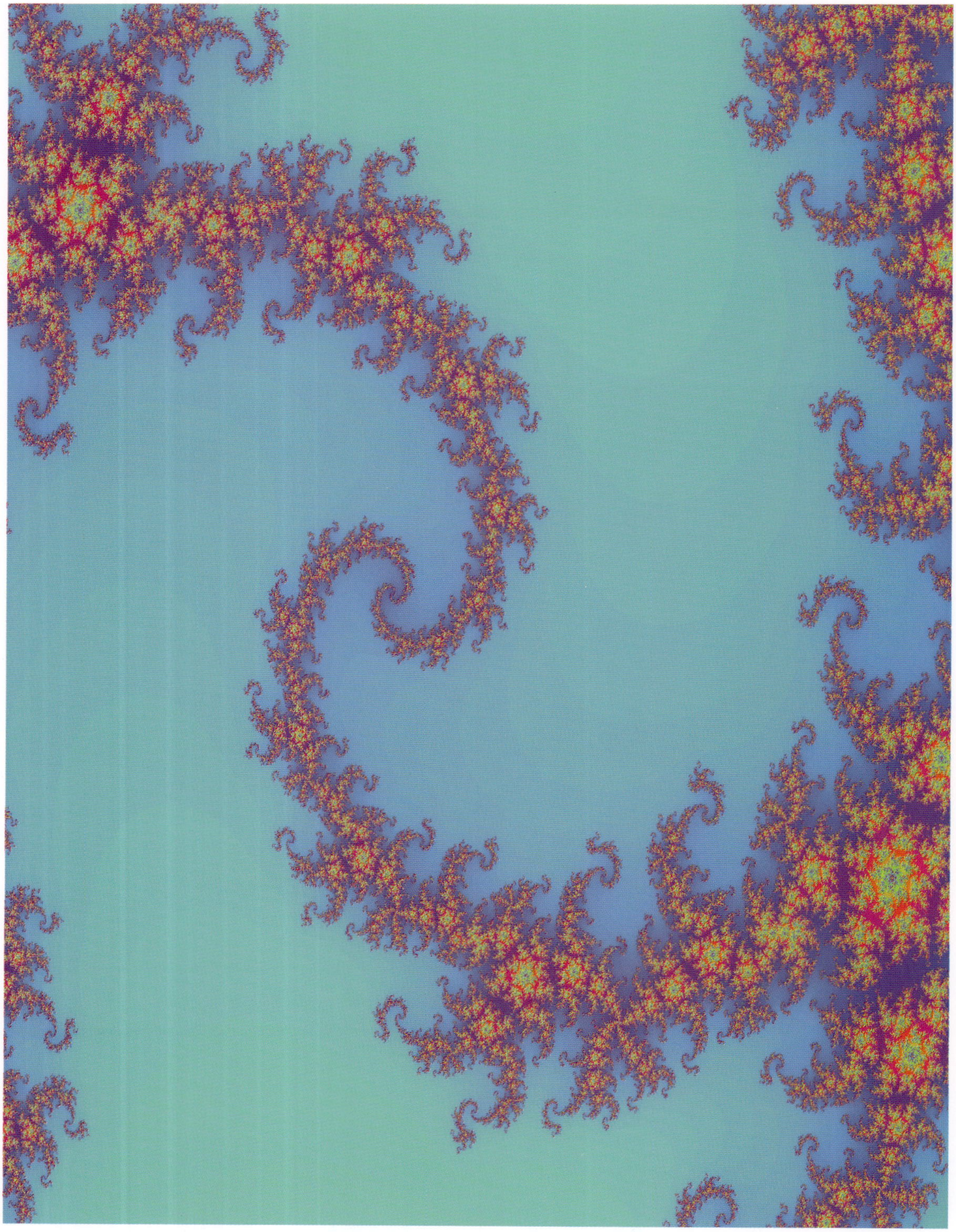

THE GOOD DEATH

Fourteen Buddhist friends
encircled him, head and feet,
right hand and left,
chanting through the days
the body's source points, nodes,
the stages of dying
into the light,

and he was our age — young!
didn't want to die; at sixty,
cancer colonizing his body,
he turned to letting go
all he loved —

meditated on entering
the world, the dust, the mist,
calling up each loved face
until it grew radiant
with light, letting life flow out
into every eye, hand, cheek, breath,
moving on to the next, the next,
the next, each rustling leaf,
the drops of falling rain,
the rain evaporating,
the sunlight
through the rain.

My friend said that afterward
they felt
his presence
for weeks.
His wife said yes;
he was everywhere,
and would not be back.

MISSING CAM

in memoriam, Camden A. Coberly, 1922–2001

We have lost him, gardener who loved
wild West Virginia hills and fomented riots
of dame's rocket, forget-me-nots, violets
in the shady corners of his Wisconsin yard.
We have lost him, engineer who crafted
wire-mesh mulch for the autumn crocus bulbs
to keep out squirrels, rabbits, and crows;
who built six-foot square canvas-stretchers
for artist son; dressers, cabinets, tables,
for daughters, granddaughters, Lenore.
We have lost him, lover of trees
whose well-trimmed hickories rained down
nuts enough for the neighborhood as he proved
that the right tool, endless patience, and a well-placed
blow could yield the nutmeats whole — or nearly so.
We have lost him, lover of stone
who polished an agate big as a dinosaur egg
for a week to bring out its burnished gold;
bearer of farmer cheese and Fetzer wine
to the writers at the close of manuscript group.
We have lost him, suddenly, in mid-stride,
Cam, who said in his fiftieth wedded year,
"I am the happiest, the luckiest of men alive," —
and even as we mourn his death
we move among the works of his hands.

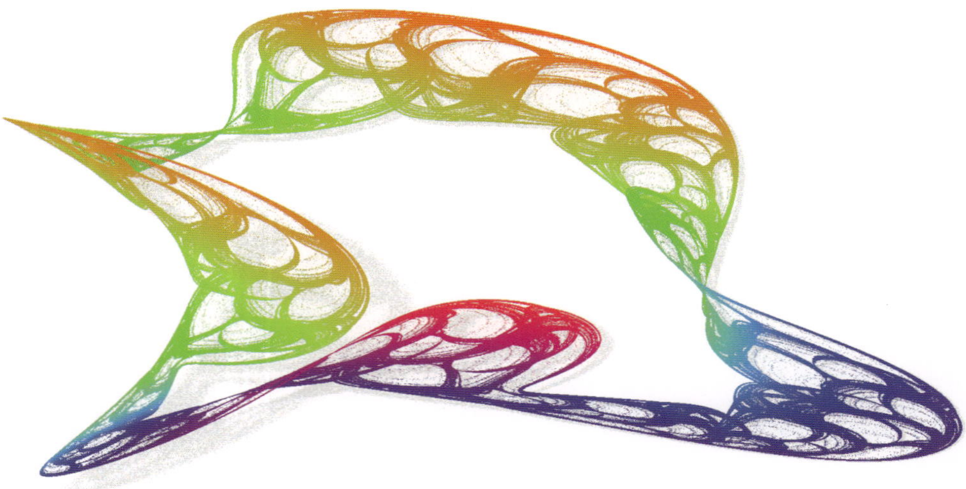

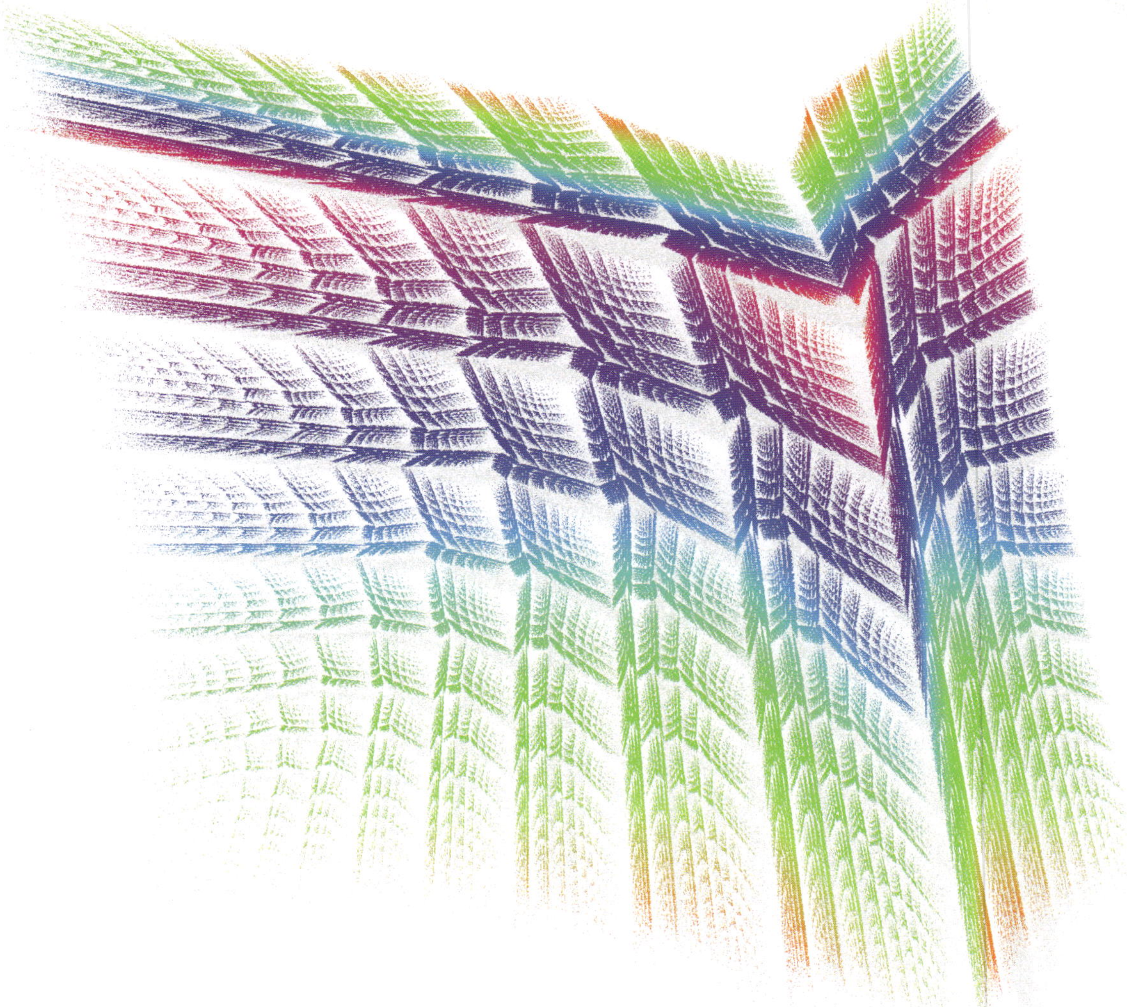

IN THE TIME THAT IS LEFT

Gulls that sleep on the bay, drifting over waves
Like a veil — the fish you picked clean today
Trail their silvery skins like shrouds,
Ghostly shapes shifting along the shore.

Why not move in sleep the way one breathes,
Keeping place in time's thrall? Turn
In another's arms, and turn again
In linked time, entrained by tidal pull?

To rise, each morning, to soft cries —
And then the hunting day.

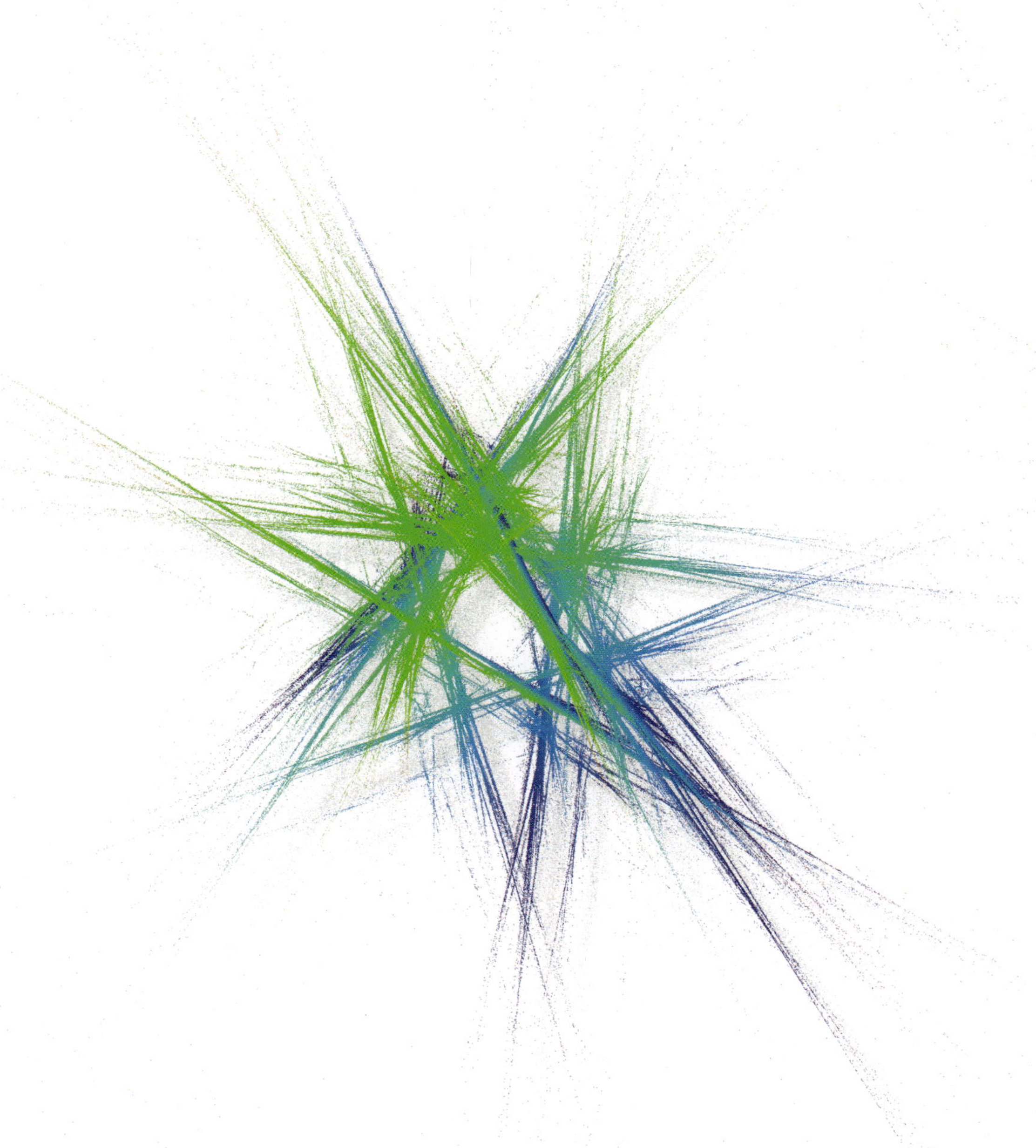

QUANTUM FORTUNE COOKIE

I've posed the question, *how to stop war?*
and clicked on Cliff Pickover's website
for my personal fortune cookie in answer,
its random pick generated by radioactive decay
of a uranium isotope, which instantly replies
to understand the nature of reality,
look within yourself.

TRAIN WRECK

Time to leave work for the night and Will,
who's punching the throttle of the electric train
on the simulated Baltimore-Philadelphia run,
revs it up to 130 for a last wild ride
as the screen blinks "Over Speed Limit!!"
and I grow increasingly agitated
at one more wreck in the virtual world
that the world has become, just buttons
and screens, till he humors me,
hits Backspace key, the emergency brake,
to bring the cars, their many passengers,
to a halt, three engine lengths from the end
of the tracks, Philadelphia skyscrapers
stacked in the background, slowing with it
my pounding heart — one less piece
of carnage, one more illusion of control.

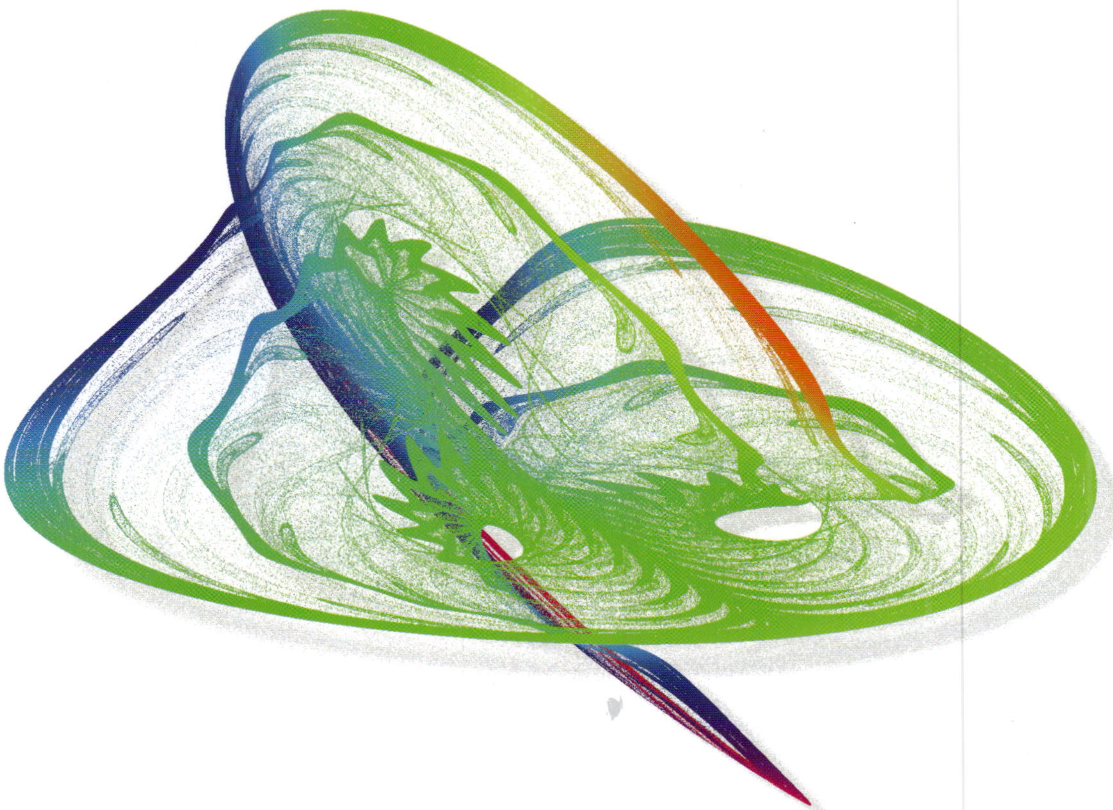

Endless Repetition (Limit Cycles)

WHEN YOU FALL INTO DREAMLESS SLEEP,

the world goes away —
color and form, the loved one
sleeping next to you, the sun, the moon,
and you — you're gone too,
free of desire and words,
your aching bunion,
your face and name —
all vanished with you —
every night you enter your bed
with perfect trust,
fall away from yourself
and all that you know
into the kind clear dark
of nothing at all.

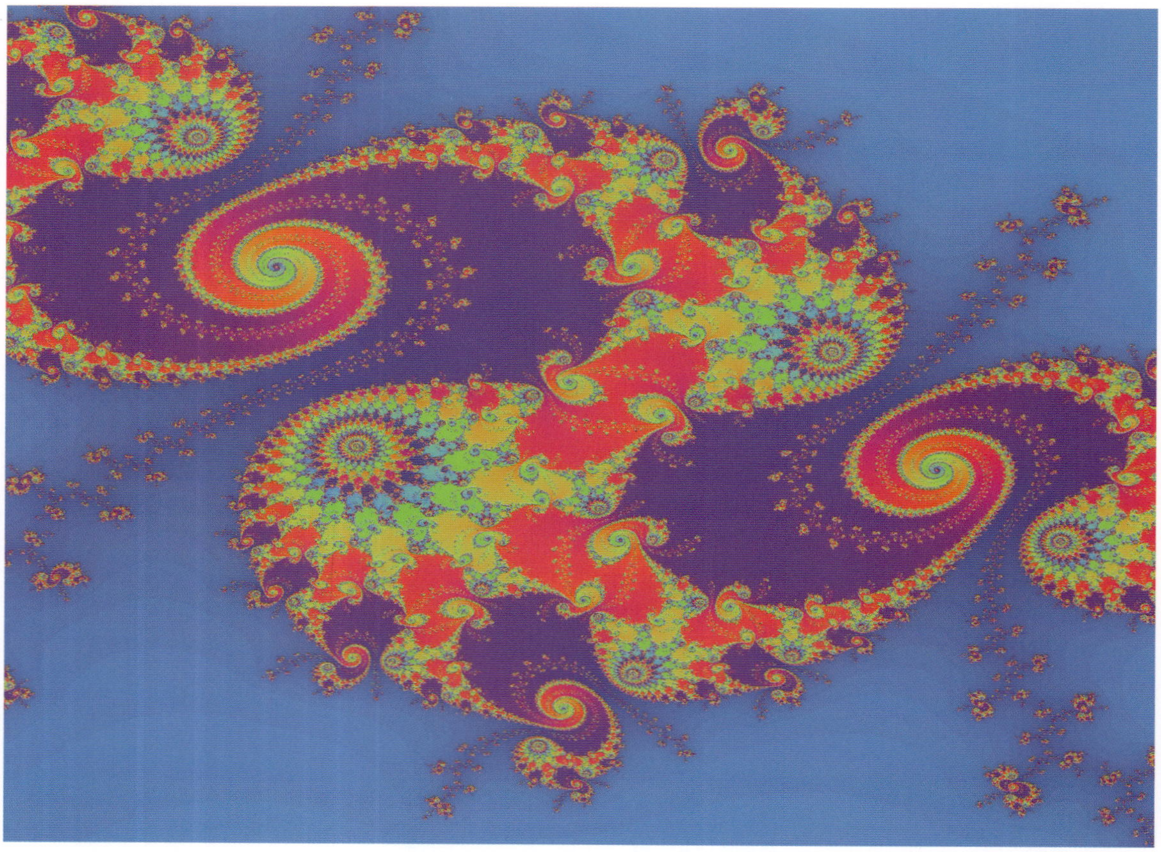

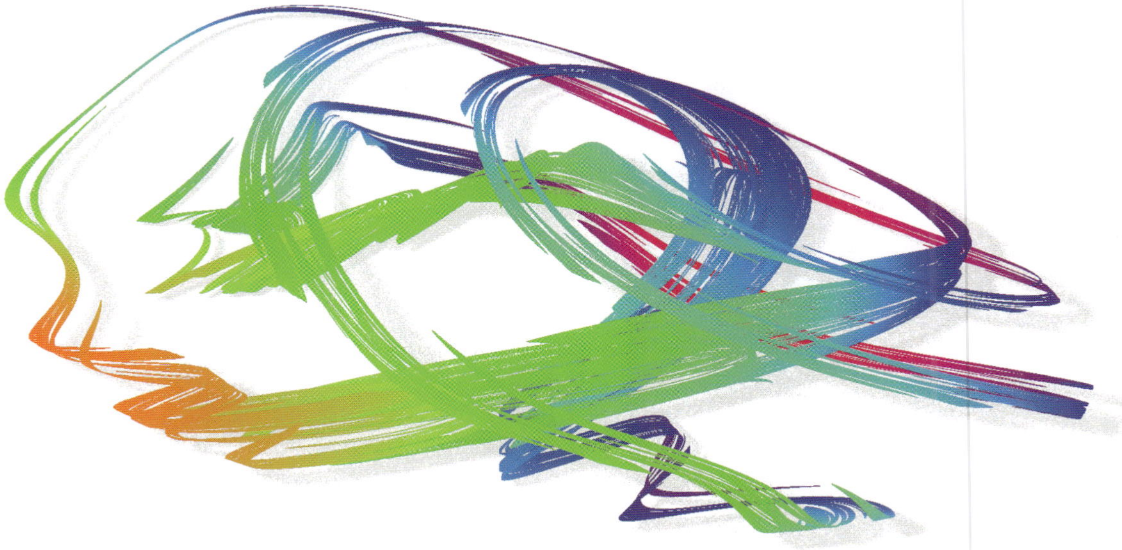

FLIGHT CAGE, ST. LOUIS ZOO

Hoops of iron rise toward the sky
And descend again, bound upright
By latticework that sieves the air.
Wind rustles the leaves of trees
Inside, ruffles our hair as we stroll
And stand and look along the redwood catwalk.
After the pair of tamarins huddled together
In an empty cage, we too huddle and laugh
At the freedom of mud and water, concealed
Heat lamp, nesting box, feeder. Wind lifts
And blows the lost feathers of flamingo and gull,
Duck and ibis, the down of city sparrows
Who've found life inside easier.

Once I saw gulls back home spread out equidistant
On a field in a great convocation of feasting.
At some unseen signal they rose and turned
And flew toward the water, circling
Clockwise and counter-clockwise
In columns over the marsh as though
The air itself were turning, the wheeling
Paths braiding and never tangling; for minutes,
They rose and fell and wove the air
In who can say what — memory, farewell, blessing?

WHAT'S THE BEAR DOING IN YOUR POEM?

The spectacled bear restlessly pacing
His concrete perimeter is looking
For a way out of this poem,
He wants no part of spectators' comments,
He knows this is no place to live,
The air not what it should be,
Cave contrived, soil thin and foreign,
He is looking for nothing he could tell you.
If he could tell you, this poem
Would be full of weeping,
And you would let him go.

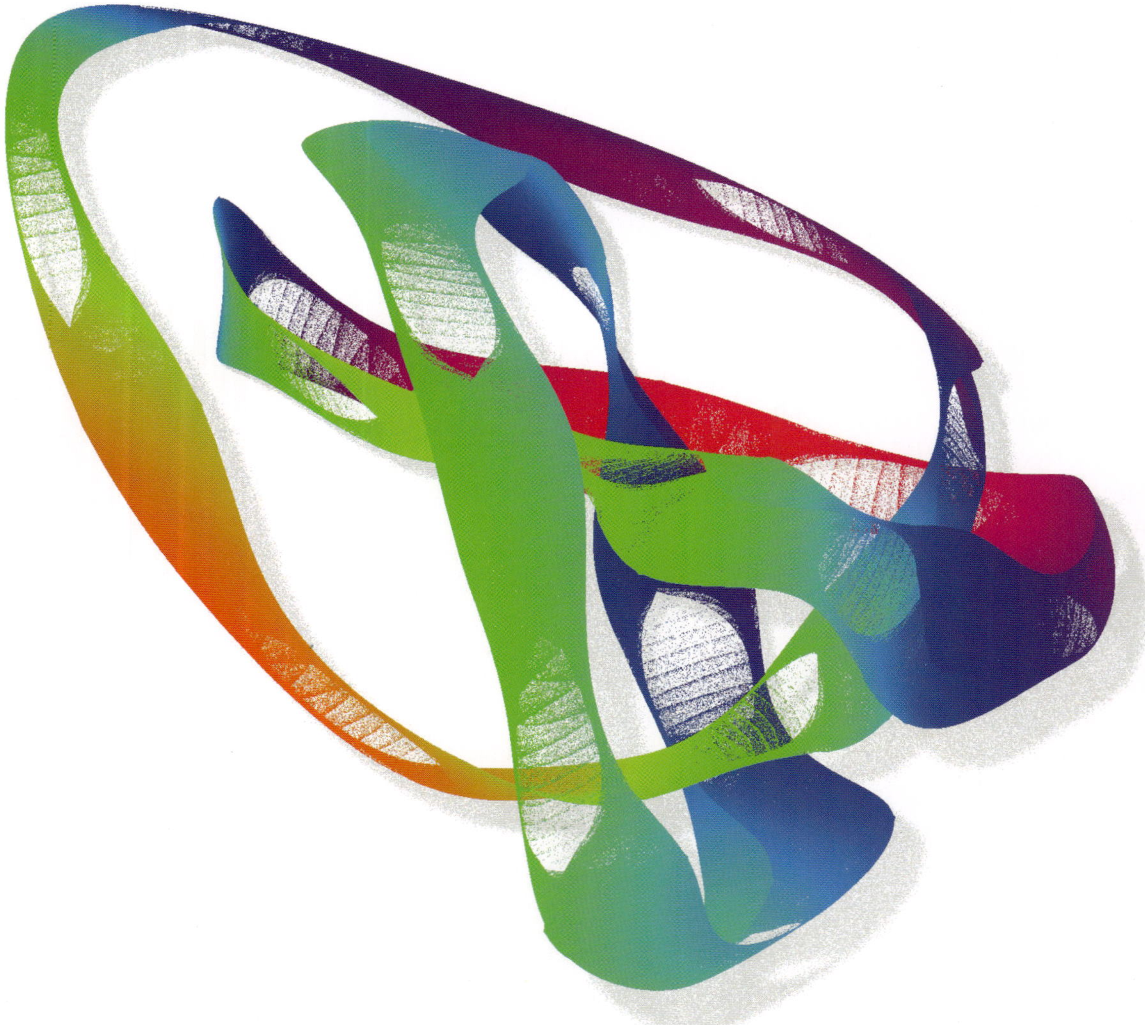

HIBERNATION

Imagine rising from your basement bed
After a sleep so long
It seems another lifetime, dreaming
Of green blueberry days, moist thimbleberries —
To struggle awake, shrug on your warm jacket,
Make your bleary way up the stairs
Out of the house
Into air sharp with snow,
Loud with cardinal song,
To take up again your life —
Stepping, hungry, into that shattering light.

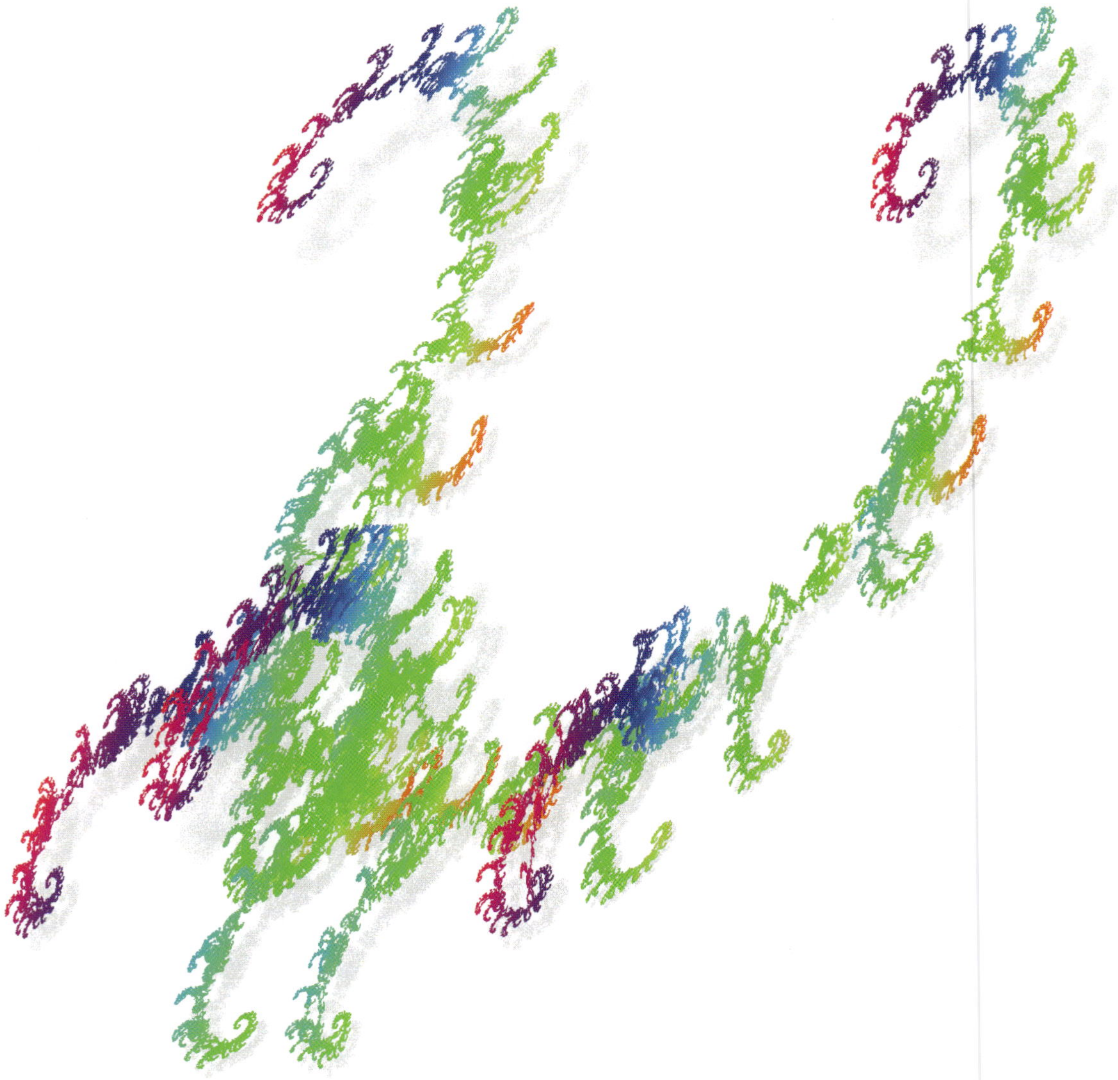

Doughnuts and Inner Tubes (Tori)

THE TORUS

Easy enough to imagine —
a donut, its surface sticky with glaze
that a fly could wander endlessly
in circles or spirals, big or small —
or our wingless selves reduced,
confined to walk the sugary world,
the air an unknown dimension,
the sweet interior another place
we cannot go and know nothing of.

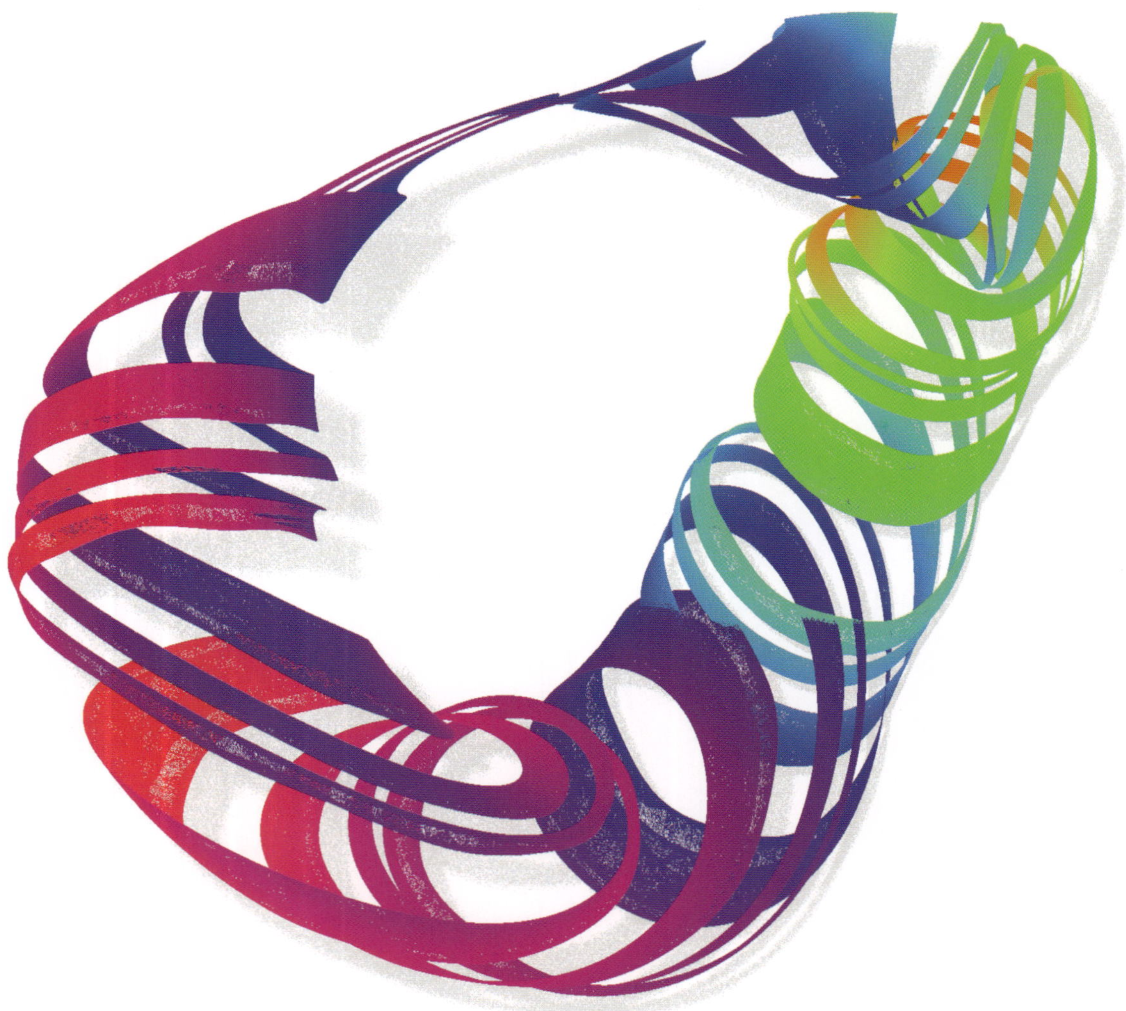

Strange Attractors (Chaos)

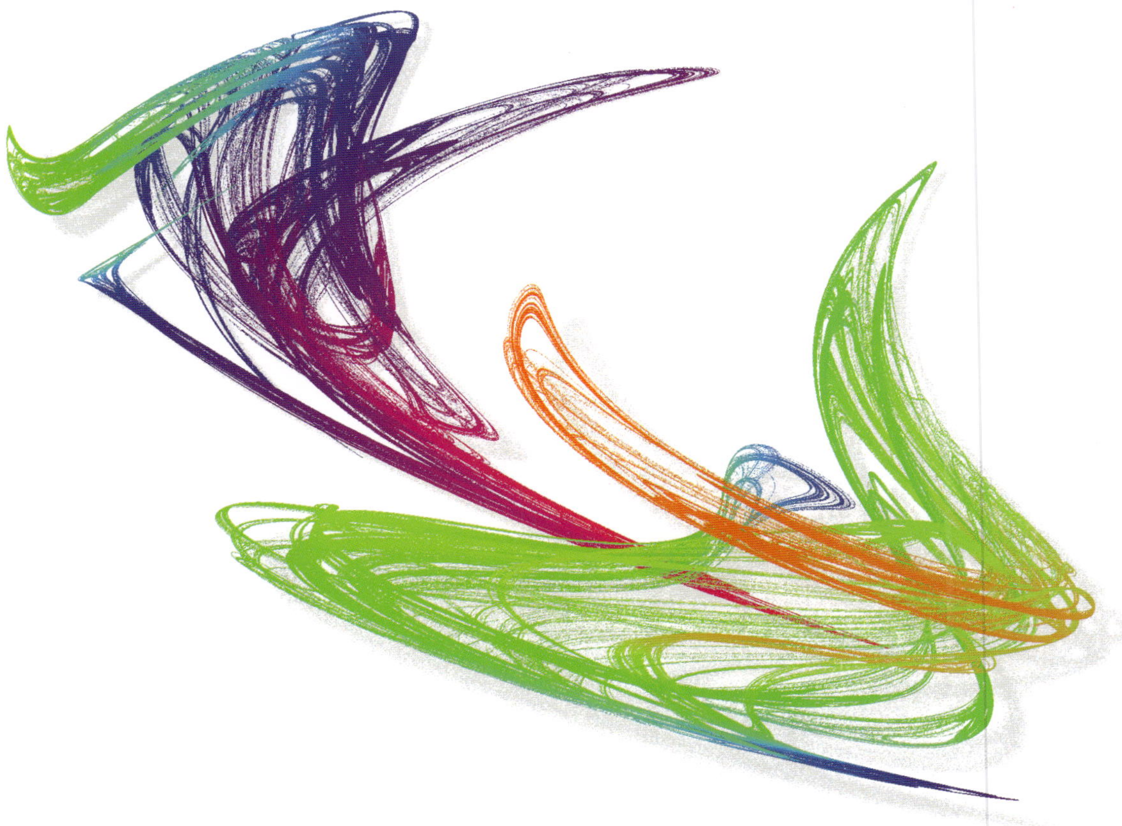

Strange attractor

Dynamical systems that exhibit chaos typically do not have orbits that visit every region of their state space as would a completely random system. Instead, the orbits are drawn to a small (in fact infinitesimally small) region of state space called a "strange attractor." It is an attractor in the sense that it attracts orbits from all initial conditions within its basin of attraction, just like a lake attracts water from everywhere in its watershed. It is "strange" because its structure is fractal, with self-similar detail at all levels of magnification. The attractor is a set of measure zero in its state space, and the orbit visits the various regions of the attractor in a deterministic but chaotic manner whose temporal sequence cannot be predicted very far into the future. Strange attractors, such as those shown in this book, have considerable aesthetic appeal, presumably because they remind us of objects in nature such as clouds or turbulent fluids whose behavior is chaotic.

A LASER LIGHT AND FOUR STACKED BILLIARD BALLS IN A CUBE OF MIRRORS

This is it, I think, as the beautiful fractal patterns unwind
On every reflected surface — red, white, black, blue,
Going on and on, each bubble containing its seed
Of the colorful pattern we stoop to see — *the secret*:
Somewhere behind our shadow-play universe, the simple
Polished shine of four billiard balls — curved space
In every direction, curving time; and our eye
Placed at the keyhole, reading the play
Of color from the east, the west, the south, and the north —
Light show lasting as long as we can watch —
This is it: best seat in the house.

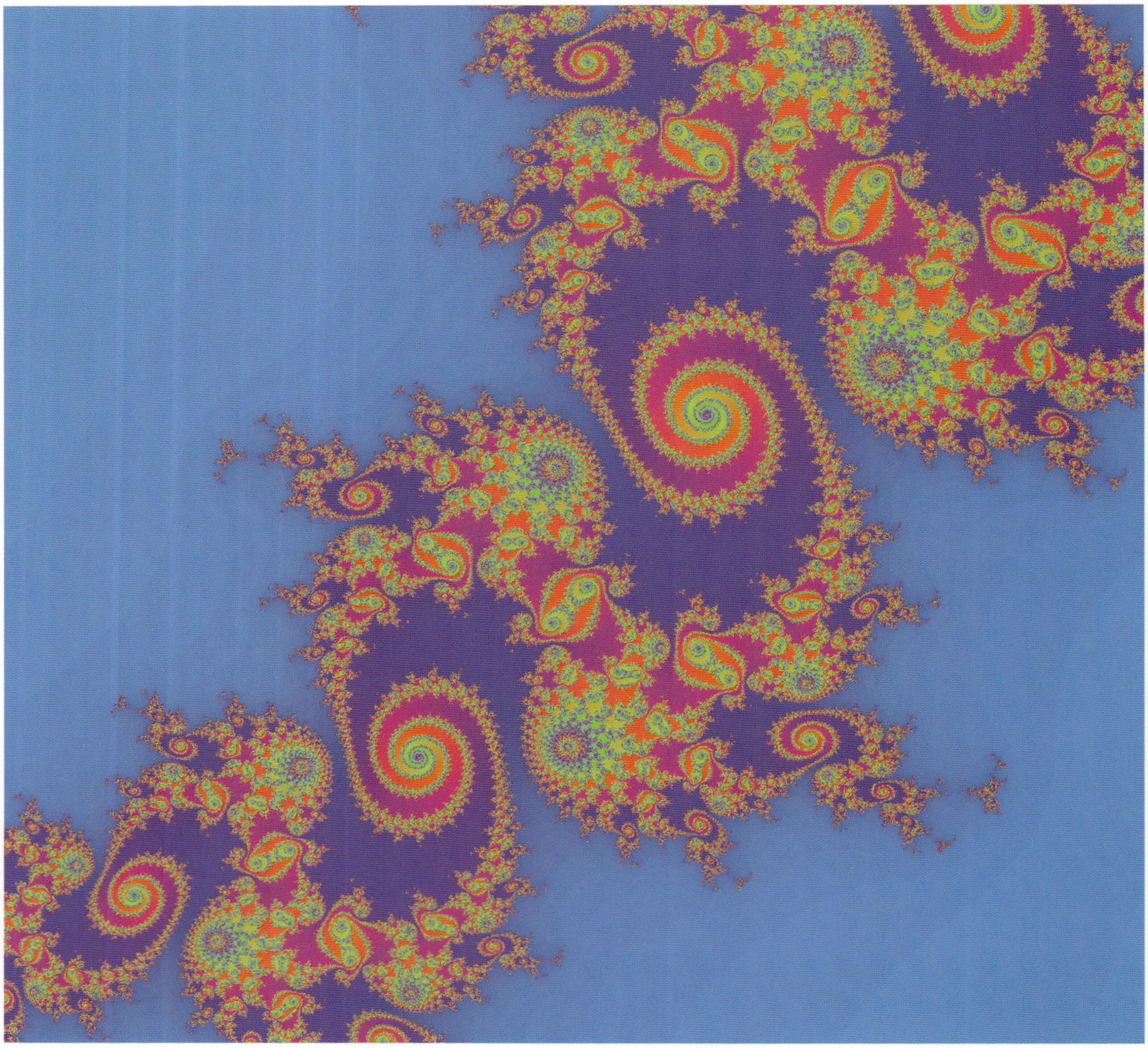

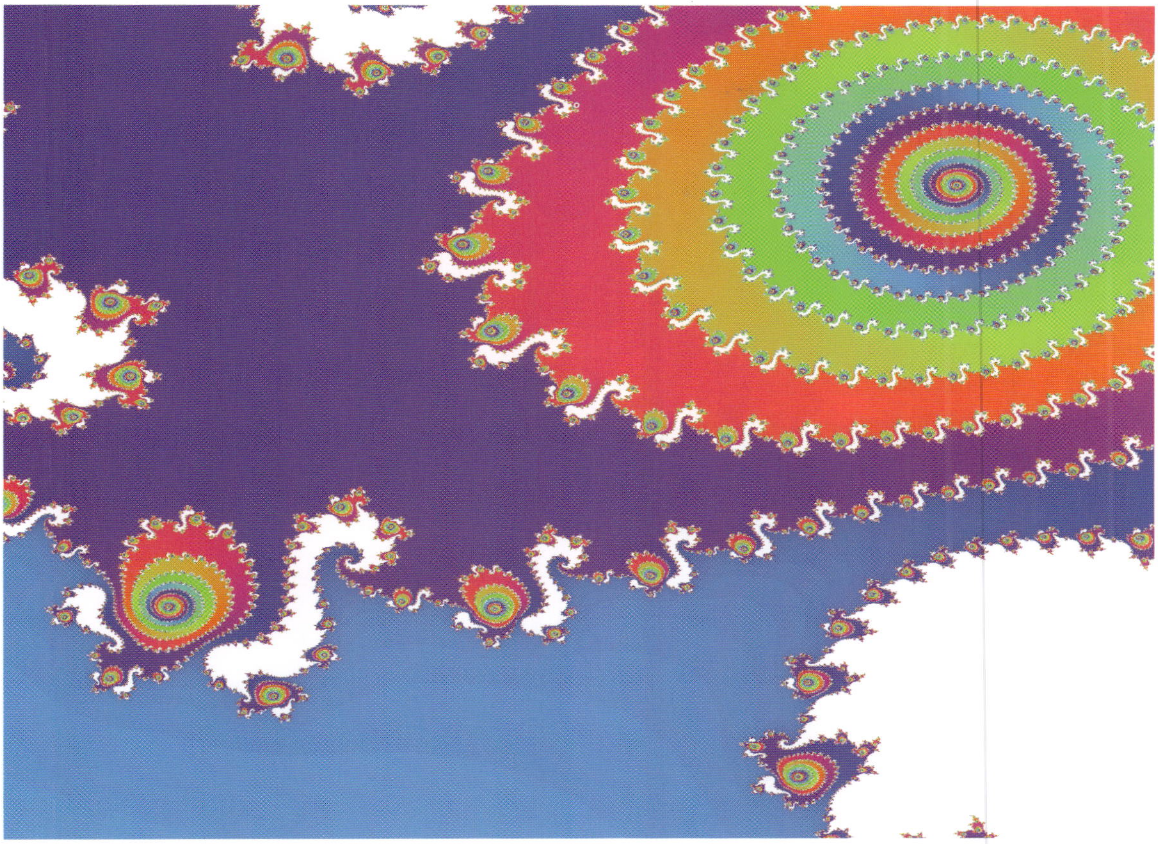

EATING A BANANA AT THE COTTAGE AND CONSIDERING EINSTEIN'S CLAIM

There is no past, present, or future, he said,
or someone said he said. I'm trying to connect
the slow time of summer in the woods,
the quick-time of waves washing the rocks,
to the no-time of the physicist — or is it that
one person's time is another's space? The years
of walking along the bluff become the trail,
the lives of coral siphoning calcium become
the bluff, the cool air in the cottage fragrant
with the banana and granola that I'm eating
with Einstein. Those banana trees already cut,
those oat fields resown as I spoon Nature's
Bakery organic mix. Those parents and children
who walked the path — shades now, or flown.

SPACE SHOT

Hurled into space travel
in its hermetic box of air,
the patented white rosebud opens
to light bulb sun, water,
cup of sterile soil, gravity's pull
loosened to micro-tug,

and the video eye
plunges its robot's nose
into the unfurling bloom.
A grid of sensing probes sniff
and digitally report the spectrum
of the space rose's microgravity perfume —

a whiff out of this world
decoded for the fashion market,
though who knows whether
the scent will signal a sweeter come-hither
to the absent bee
or the compound chemistry of loneliness,
of separation from the earth.

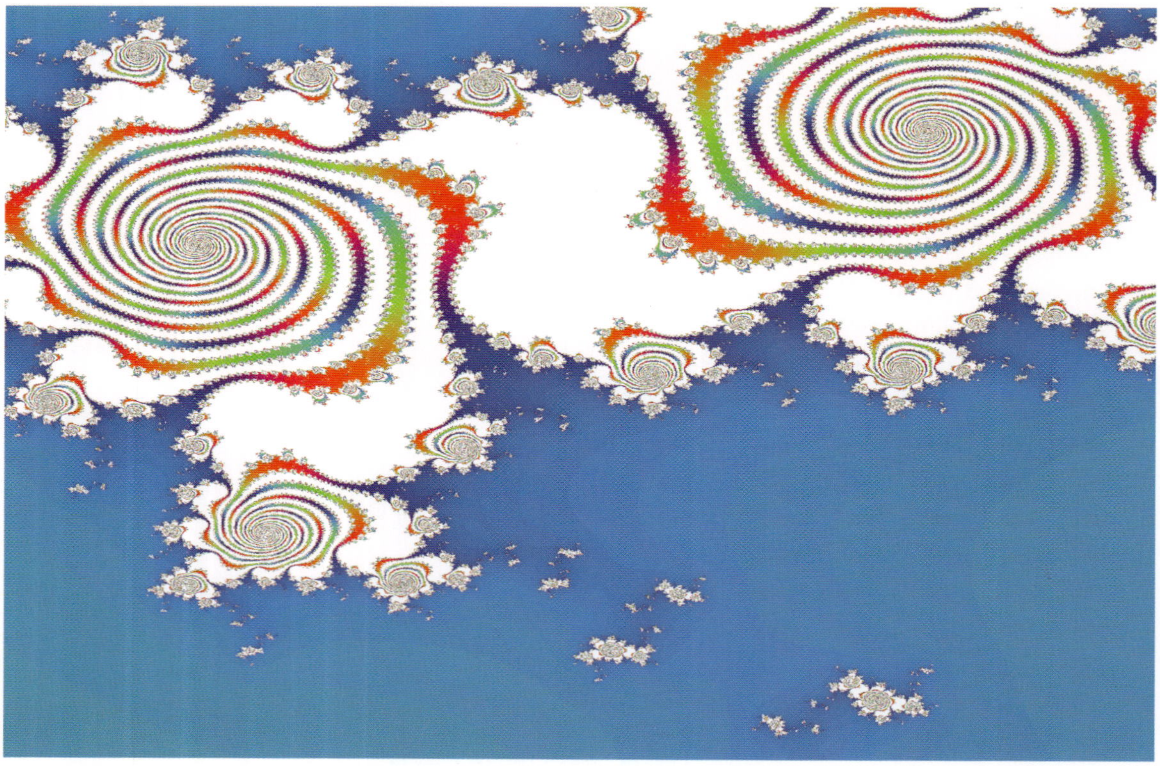

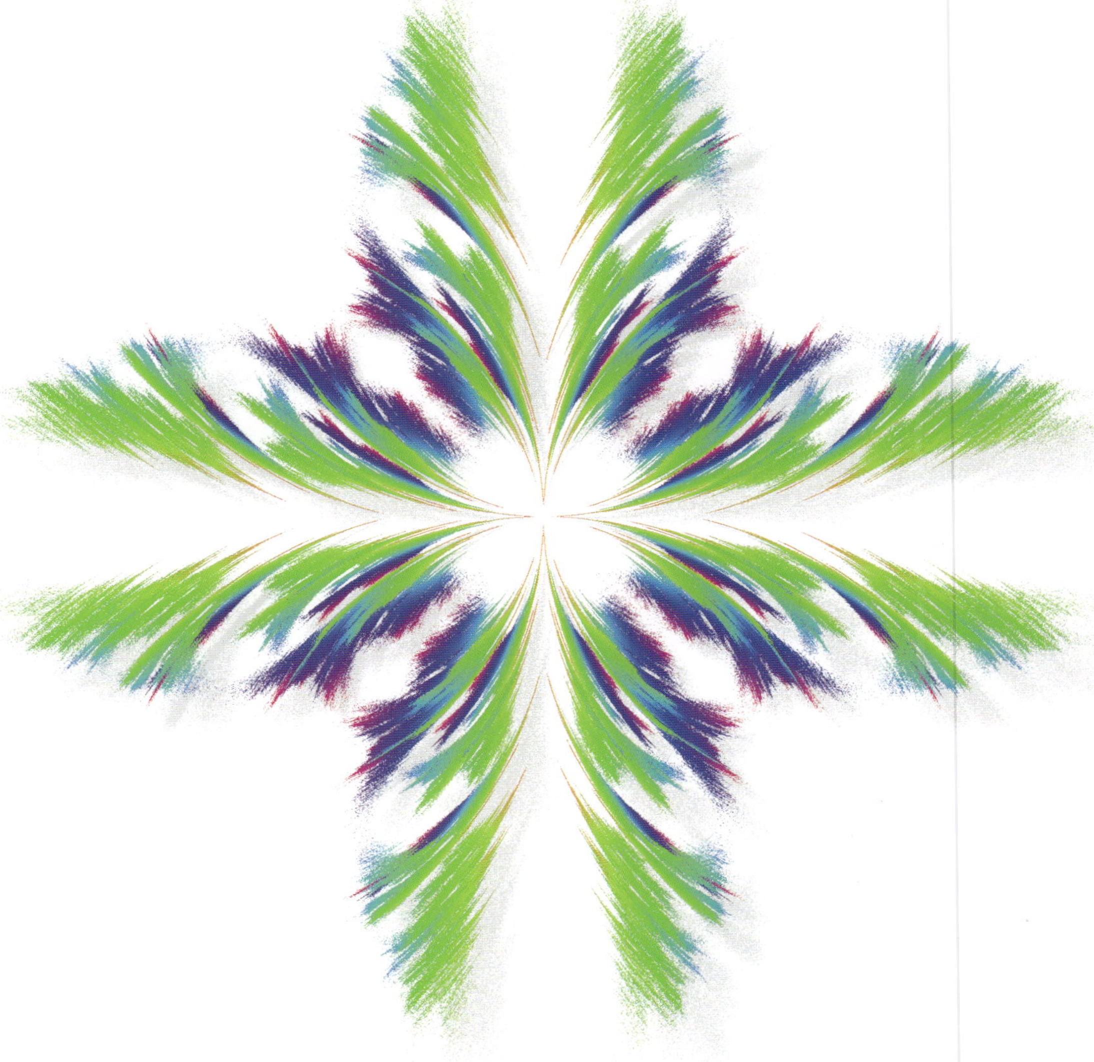

Chapter 4.

Routes to Chaos

One of the advantages of being disorderly is that one is constantly making exciting discoveries. — A.A. Milne

Forks in the Road (Bifurcations)

Bifurcation

When a control parameter of a dynamical system is slowly changed, there are values of the parameter for which the behavior of the system changes abruptly. For example, the vibrations of a musical instrument constitute a dynamical system, and the flow of air in a wind instrument might be a control parameter. When the flow exceeds a critical value, the sound might suddenly change its character. The behavior might change from periodic to chaotic. In a mathematical system the solution might suddenly become unbounded, with the value of one or more of the variables growing without limit. Such points are called "bifurcations" or "bifurcation points." The existence of bifurcations is evidence that the system is deterministic, and it provides a way to distinguish chaos from randomness, since a random system usually does not bifurcate in this way. There are dozens of different types of bifurcations, and they represent an active area of current research.

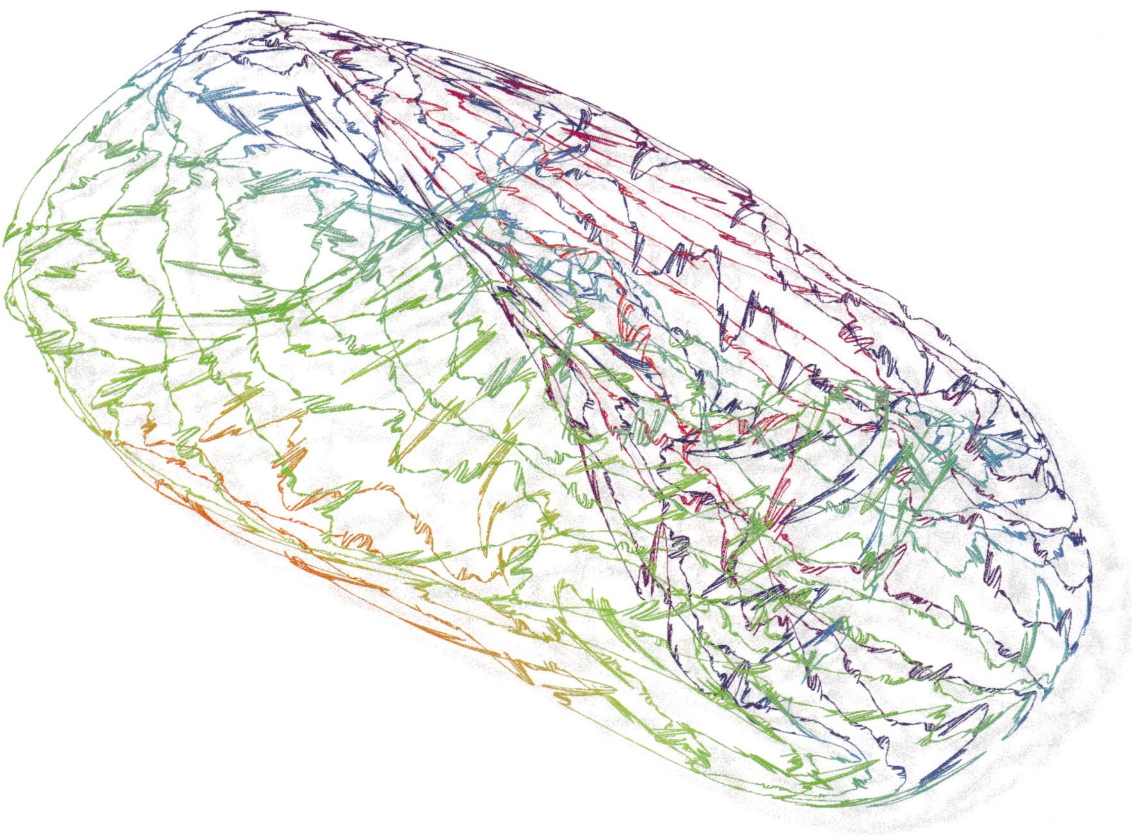

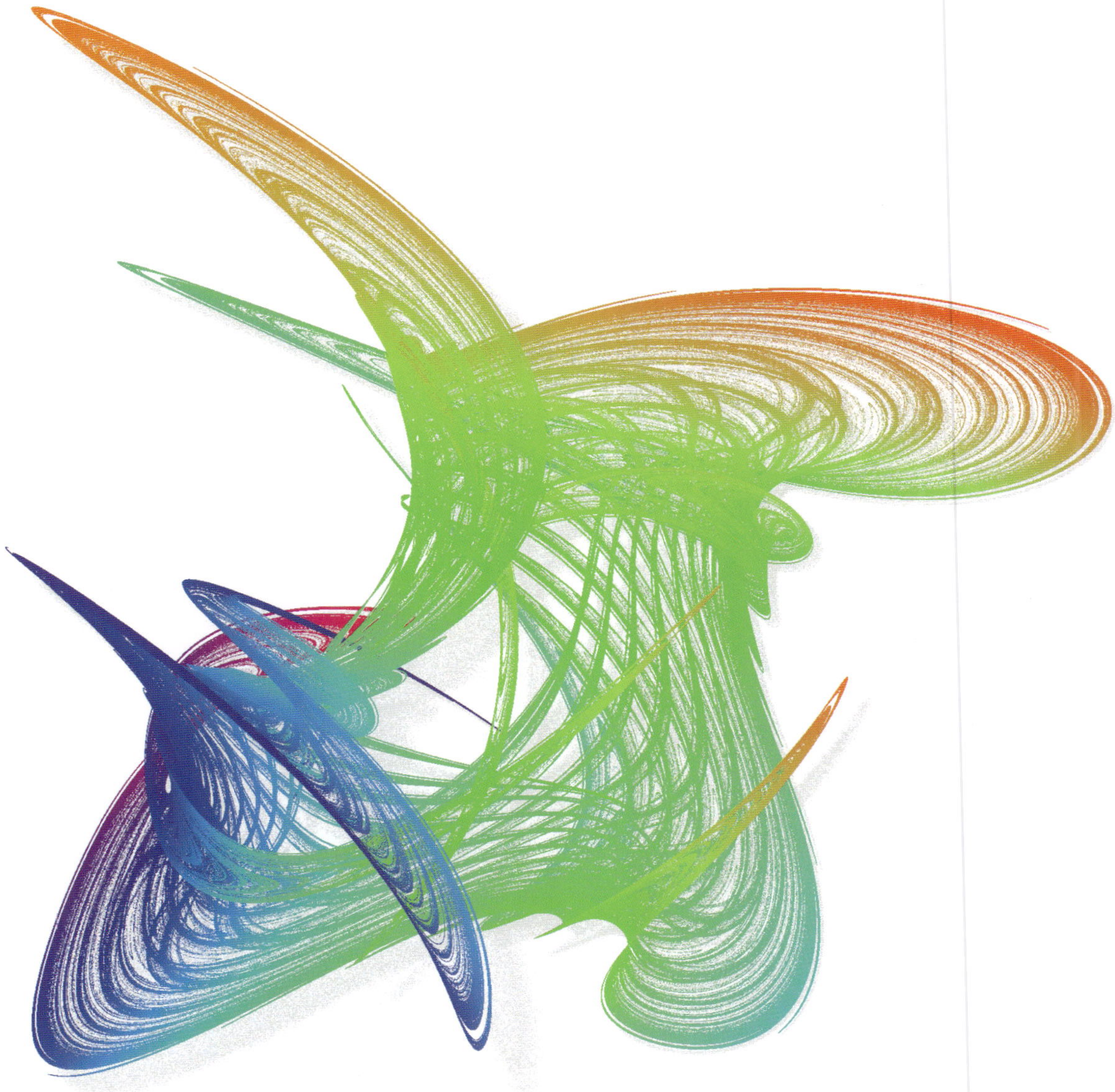

BIFURCATIONS

This is the path that pitchforks
in the yellow wood — the one
where you wanted to travel both,
science and poetry,
physics and art,
and so bounced unpredictably back and forth,
taking each as far as you could.

MIGRATION

All day the gulls and the blackbirds
Have gathered, wing streaks in rain;
I wonder what they say as they chatter,
Facing each other in darkened assembly
Under the clouds' gray shawls.
What would we say, gathering to leave
Our cities and towns, taking only
Our language with us, our maps
Of the night sky as the cold bore down?

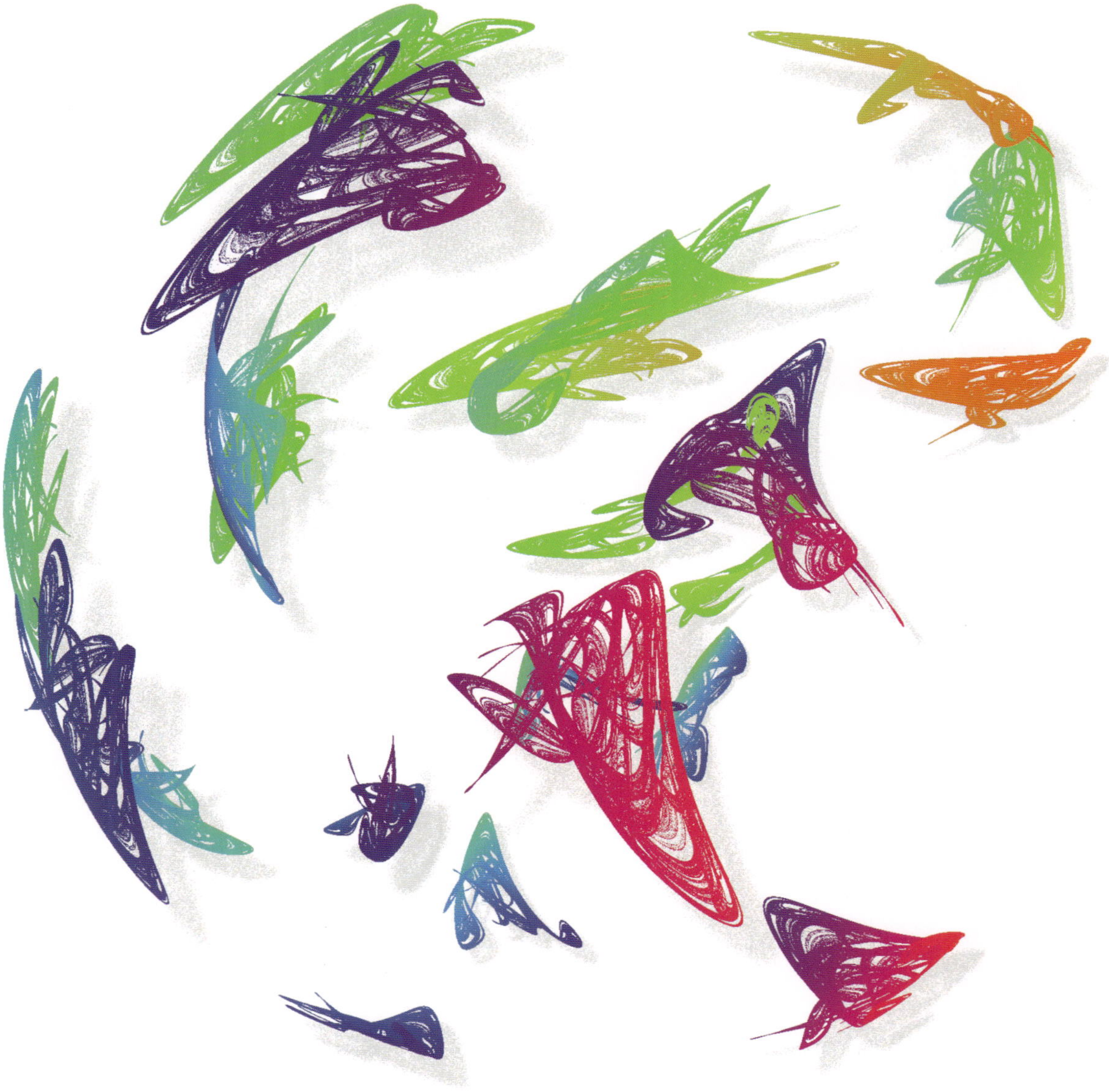

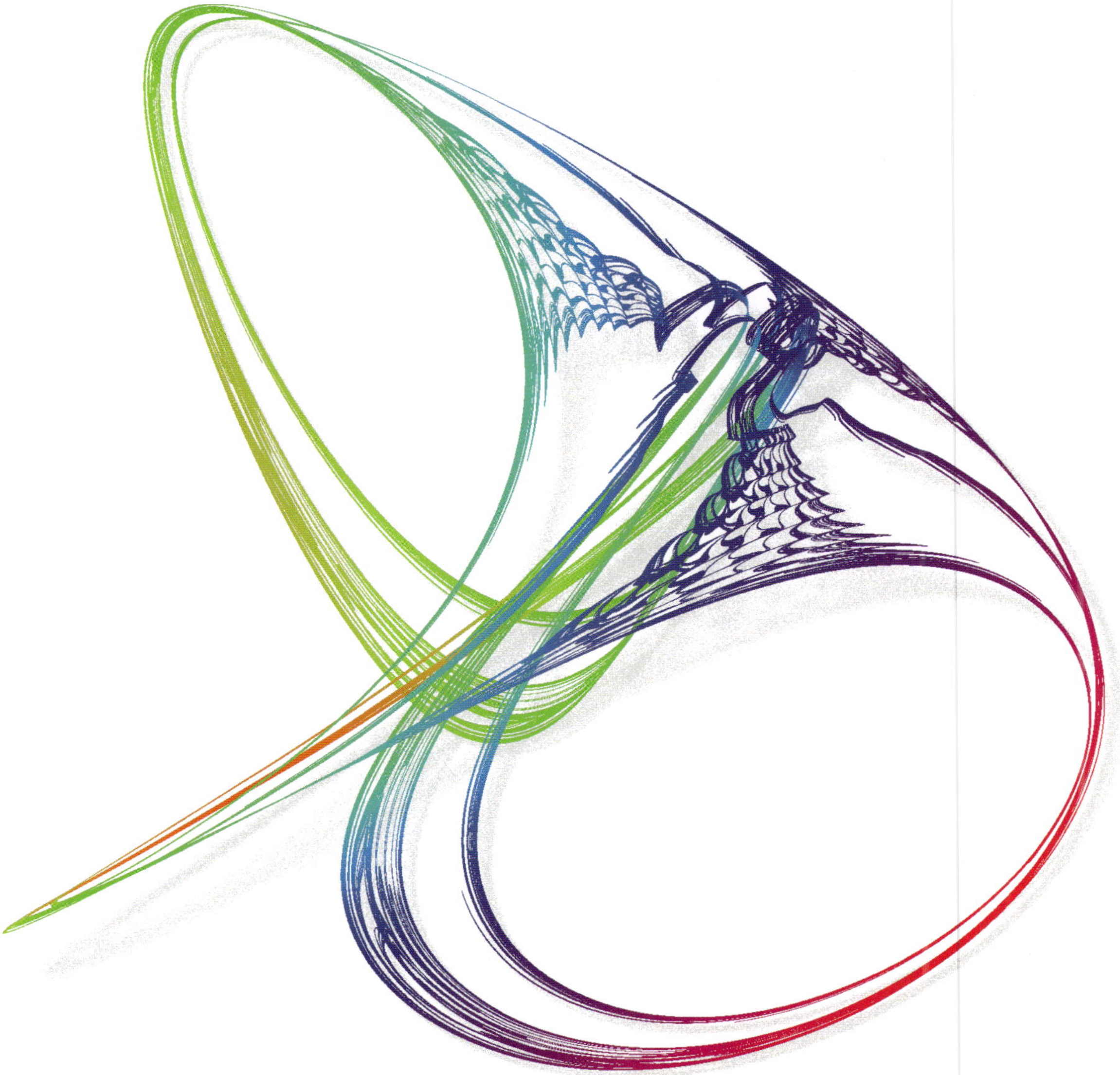

Phase transition

A phase transition is an abrupt change in the characteristics (or bifurcation) of a complex system when a control parameter passes through a critical value. A common example is the freezing of water when the temperature is lowered to 0°C, or its boiling when the temperature is raised to 100°C at atmospheric pressure. Near such phase transitions, large fluctuations in the variables are often observed, and the system is very sensitive to small perturbations, with the effect of the perturbation propagating rapidly throughout the system.

FIRST FROST

Full moon and the cold
Hurried me along the night round
With the dog, whose nose
Was as slow as usual, exploring
The ground. I thought he had found
Glowworms, or swarms of fireflies
Just crawling out,
Till I saw that the grass
And the dog were drenched in drops,
And every drop a cup
Giving back the moon's light.
The next day the first maple leaves
Waved goodbye from the grass.

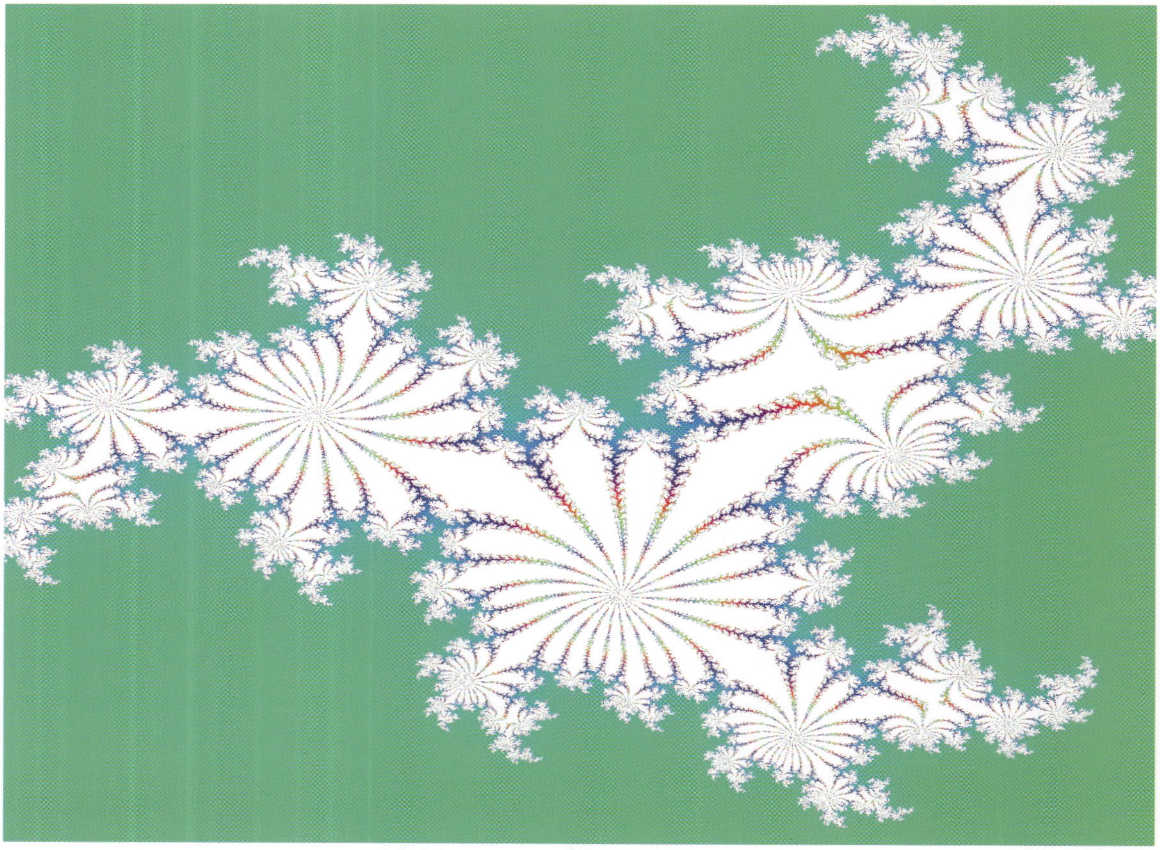

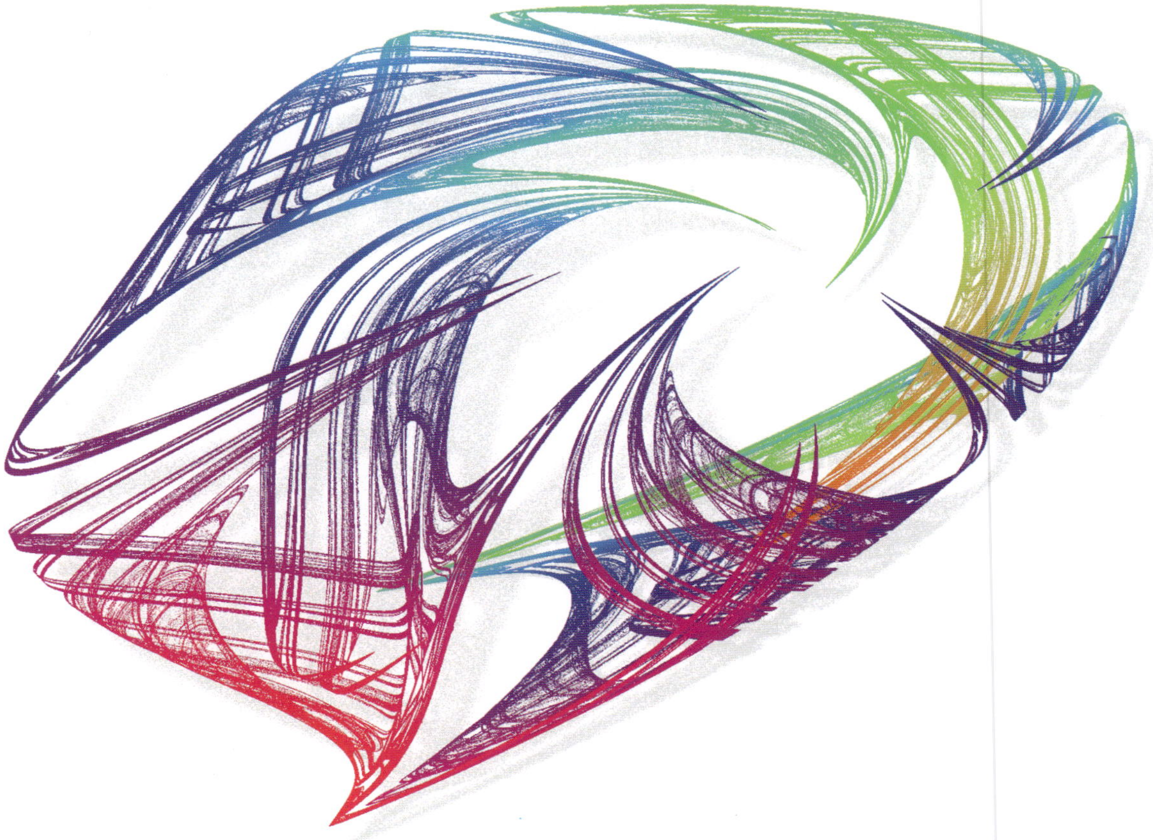

WHAT WE DON'T NOTICE IN WINTER

Is the way, at sunrise,
The shadows of lodgepole pines
Stripe the mountain face like bars,
Reduce to pools of dark at noon,
Or with the low sun's passage down,
Swing open the other way,
Gate to nighttime stars, the gibbous moon —

Or how, in brief transit over snow and rock,
That interrupted light,
Sharp-edged or softened, reaps
In its dark scythe
Every color of the sunlit world
In shifting, complementary hue —
From sun, the green; from reddish bark, the blue.

12/12 MADISON

Dear Ones — 9 degrees last night
and now the fog, rising with
the temperature from our ten inches
of snow, rimes the trees, the grasses,
making a white-furred world
through which we dash into our buildings,
our hectic holiday lives — and I want
to slow down, let my lists go,
cancel meetings, forget the tasks,
go back out into the quiet of it
on my whispering skis, past marsh,
grass ferny with frost, past
Picnic Point's hoarfrosted trees,
to watch the water rising into cloud
and crystal, my breath making white
feathers of my eyelashes, cheeks warm
with heart's warm blood, moving into
this morning like no other —
I want to say, when I come
to the end of my days,
that I did not forget
the beautiful fragility of the world;
that there were mornings I entered
into its brief, altered, and altering
weather–entered, and came back changed.

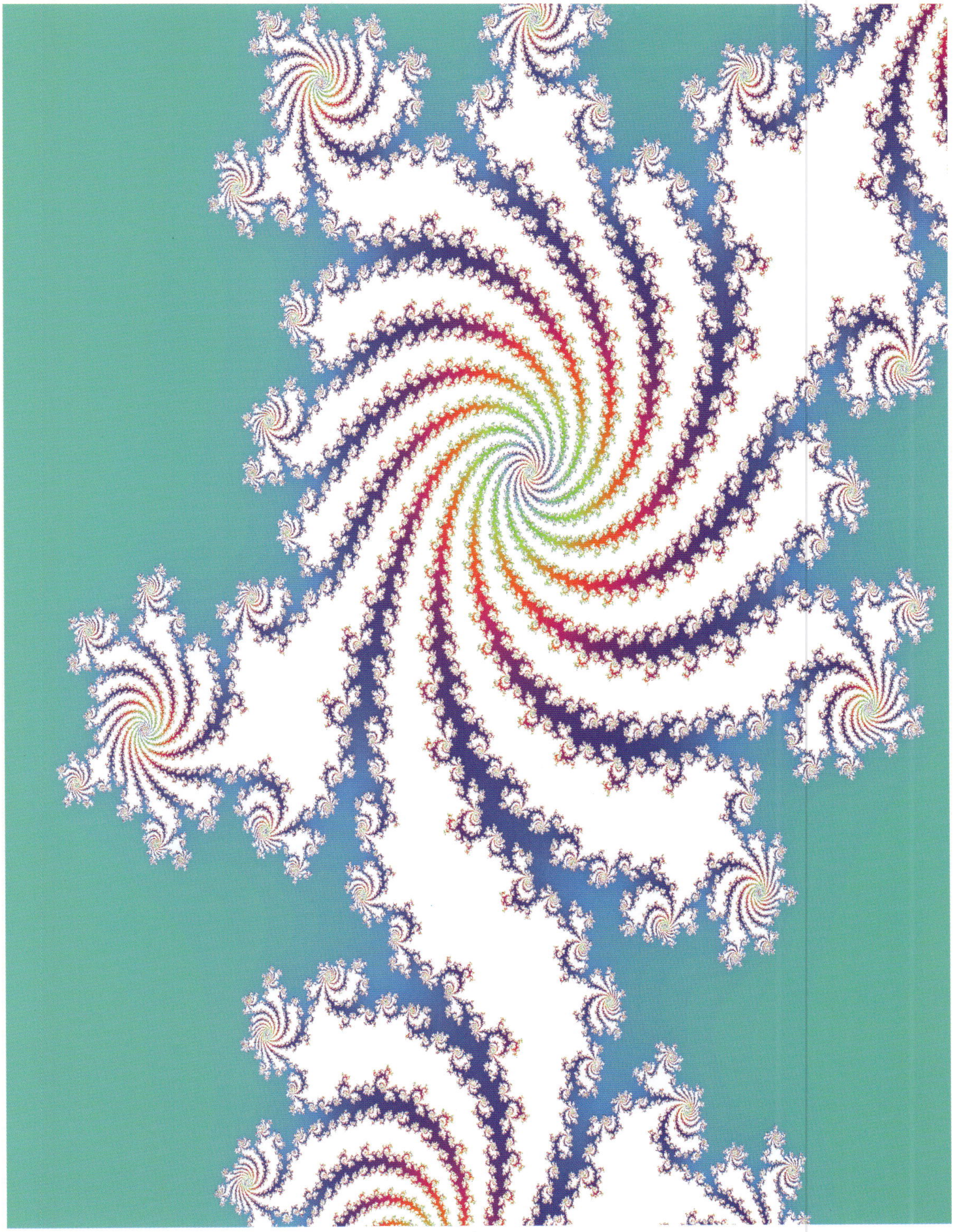

12/13
MADISON

Dear Ones — fog all day, freezing
to hoarfrost flowers in the tall grass.
White votive cedars surround the pond,
all their branches curving up.
Willows trail their iced gold stems.
At noon the backyard rabbit came
to the sunflower seeds, her whiskers
feathered in visible breath. So mild,
this hovering between one season
and the next. I drive through mist
to the almost-empty garden center,
carry away the last 200 late-flowering
tulip bulbs half-price — pinks, reds, yellows —
plant my wedding bouquet for May.

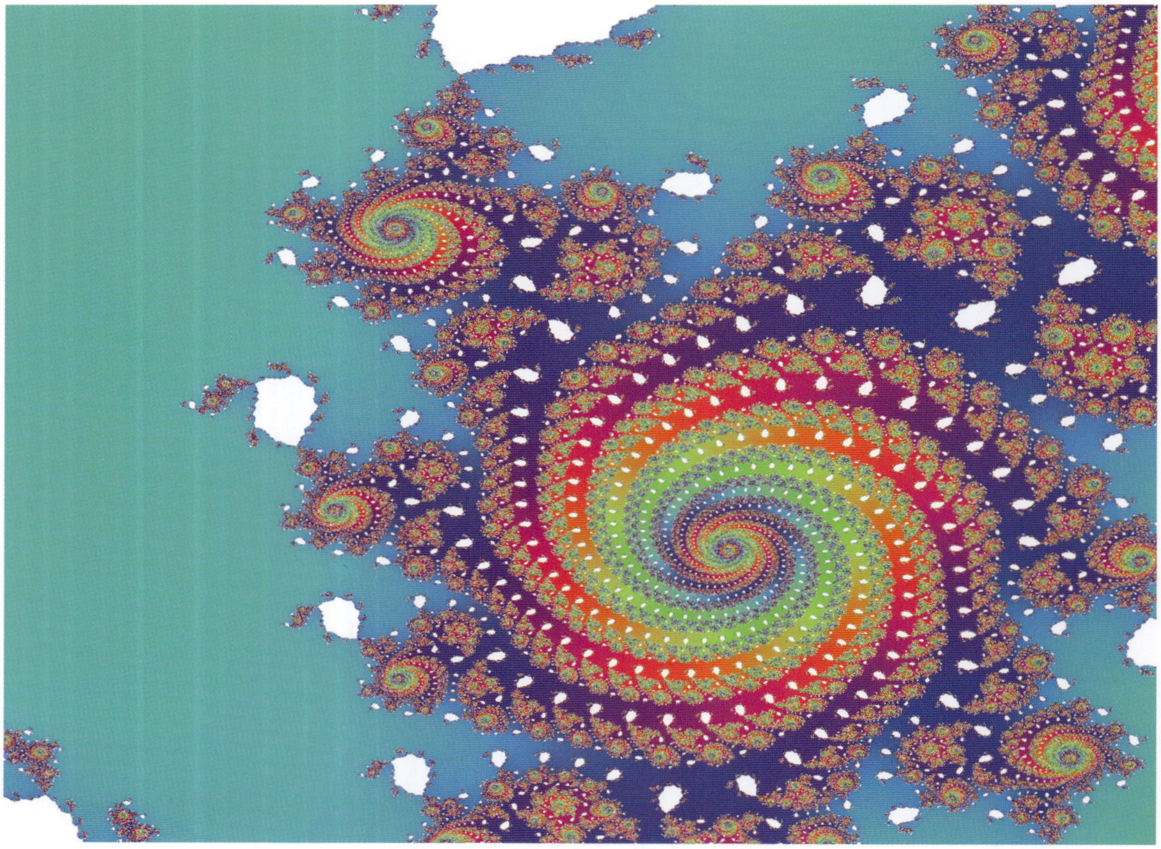

YOU LEAVE, AND I MISS YOU

for Will

I asked for this time, this space
ringed by cougars and mountains
so that I could write; though it's you

I think of, your long-limbed arms
folding around me, the set of your lips
and jaw as you call up ear-melodies,

The Irish reels and Swedish hambos
that pour through your fingers
on the accordion's buttons and bellows.

Across a roomful of folk singers
we are dancing with each other,
singing and swaying to *Roseville Fair*.

How clearly I can see and follow
your body's every gesture
though I am here and you are there,

Though outside is the dusk that will require
presence, attention, from all who move
through these cougar-hunted woods,

The wolf pack a noose drawn around us,
the coyotes threading their way,
the elk browsing uneasily toward town.

FLOOD

The marsh spreads, reclaiming
The rolled soccer fields, the swallows'
Bridge, the parking lots. Meters
Float in the silvery water,
A stiff new breed of duck.
The path we walked
We could paddle now, and still
The rains come, teaching us
The ways of water seeking its level,
Going straight for the heart
Of earth. Downstream, the Mississippi
Cuts a new channel.

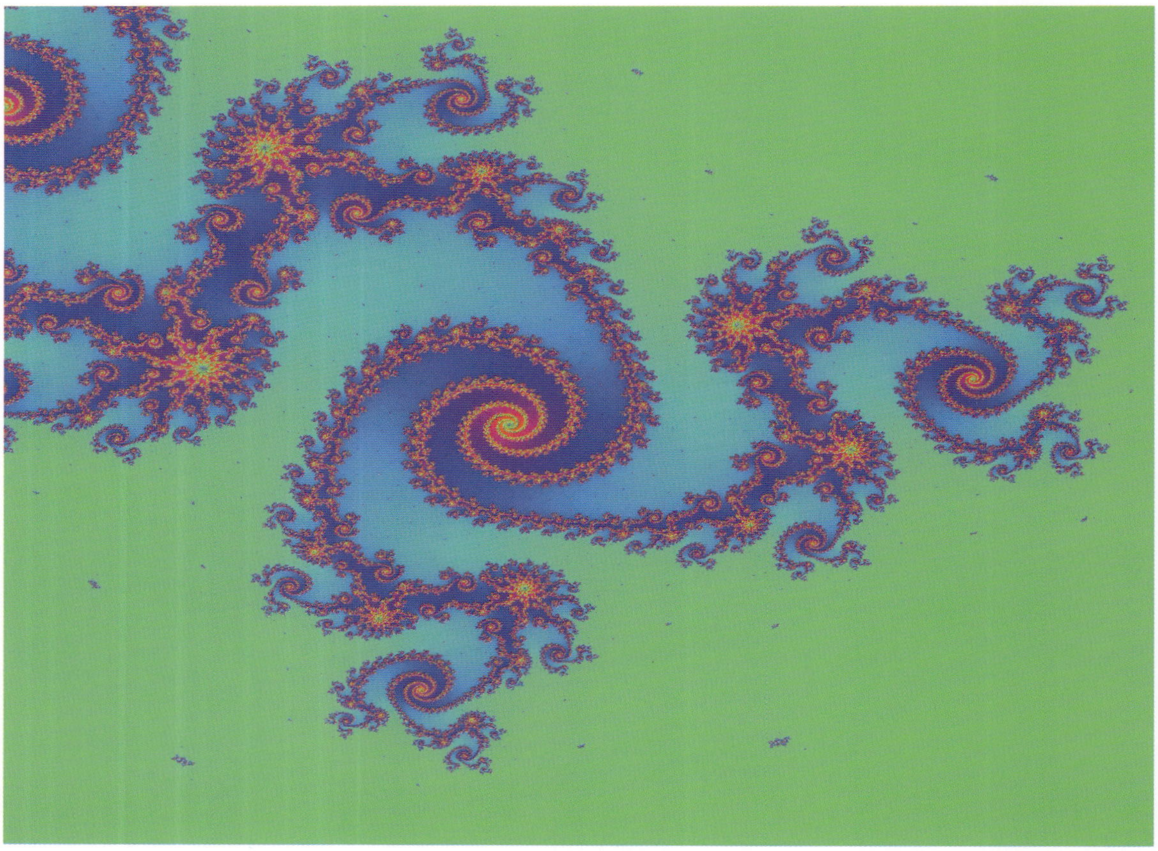

Dimension

Dimension is a familiar concept for dynamical systems that evolve in real space. A train moving along a railroad track is an example of one-dimensional motion, even if the track is curved. An ant walking across a sheet of paper exhibits two-dimensional motion even when the paper is not flat. A fly buzzing around the room moves in three dimensions. When the variables are more abstract, you can think of them as the axes of a state space whose dimension is equal to the number of variables and thus can be arbitrarily high. It is hard to visualize spaces of more than three dimensions, but they pose no mathematical difficulty. There are even situations in which the dimension is infinite. To accurately model the weather down to the smallest scale in space, you must measure the variables such as temperature and humidity at infinitely many spatial positions, each one of which corresponds to a new axis in this infinite-dimensional state space. To specify completely the behavior of a fluid such as a turbulent river requires six variables, three for the position and three for the velocity, for each molecule of water in the river.

THE THERMODYNAMICS OF WATER

Slash it with a knife —
it heals as fast as you cut,
and cools the blade.

Let it flow —
it can slice rock
like a knife.

Lock it up as glacial ice —
it flows, river of blue light
that you can slice.

Set it upon the fire —
it boils and steams and vanishes,
returns as rain

Watering the soil —
seeds split their husks,
push into light.

MOVING IN

for Will

I was used to solitude and so were you;
I'd filled up my house with thirty years,
and you had too in a town an hour away,
so we took it slow, your arms
full of ripe tomatoes and dancing shoes,
space for your car in my garage, a spring
wedding, a toothbrush glass.

Weekly you arrive: first, canoes,
the sixteen-foot solo for fast rivers,
the shallow-bottomed fiberglass
for Boundary Waters lakes,
the light swift Kevlar for portage ease,
the old aluminum, for nostalgia;
and kayaks, four: sea-faring solo
from the Baja coastline trips,
whitewater solo from Wisconsin's rivers,
folding solo from canyon-cutting meanders,
folding tandem from Everglades trips;
you haul a ghostly fleet to float
above our heads in the garage
by new ropes and pulleys, dripping
sand and salt and pine duff.
I empty out a closet, three dresser drawers,
make space for maps.

Then all the tents and sleeping bags arrive,
trunk-loads of firewood cut and split
through twenty winters to stack in the garage,
bringing the scents of elm and oak
and the opossum who prowled your shed;
heat leaps up in the Greenbriar stove.
Your Cajun accordion glitters from the shelf;
your Irish accordion stands beside the hearth.
I repaper the kitchen cabinets, send
boxes of fondue pots and chafing dishes,
shish-kebob spears to the Good Will depot.

You pull in with your compost bin
and humus, buckets of river stones
tumbled smooth in the Green, the San Juan,
the Little Vermilion, pulled from the corners
of your rooms; I fill shallow baskets
to the brim, begin to clean the freezer out.
You arrive with ten years of frozen grapes,
Fredonia, Niagara, Lindsley; and Sweetpea,
your tabby cat, with litter pan, feeding bowl,
and signs that warn *New Cat Being Introduced:*
Do Not Open Door. Awful Catfight Will Occur.
My Ava sulks and growls, spits and hisses;
we keep them on separate floors, till the week
we accidently shut them both upstairs,
find them walking peaceably about.
Now it's a lap for each beside the fire.

We roll your picnic table through the ice,
unpack your records of the Everly Brothers,
Cajun accordion, Norwegian fiddling,
two hundred operas. I sign up for house
repair, new windows, order perennials.
You line the basement wall with shelves,
fill them with mason jars and nails, contract
for shingles and five inches of roof insulation,
chip off ice dams, explore the crawl space
in your caving gear and headlamp,
put the *For Sale* sign in front of your house.

And now the flood of couches, Mission
lamps, Cuban oils, Lake Michigan pastels.
A white refrigerator replaces avocado,
all my spaces suddenly new. I'm playing
Buddy Holly, Puccini, Verdi on your stereo.

Routes to chaos

In a simple deterministic nonlinear dynamical system, chaos typically occurs only when the control parameters are adjusted carefully to appropriate values. Other values of the parameters usually lead to a static equilibrium or perhaps to a periodic oscillation or unbounded growth. It is interesting to ask how the transition from regular to chaotic behavior occurs. A common route to chaos is by way of period doubling in which the period successively becomes longer by a factor of 2, 4, 8, 16, and so forth, until it is eventually infinite. Other routes have also been observed such as intermittency, in which a periodic signal is interrupted by short bursts of chaos that become longer and more frequent as the fully chaotic regime is approached. Knowing the routes to chaos helps one recognize determinism and to identify when a system is on the verge of becoming chaotic.

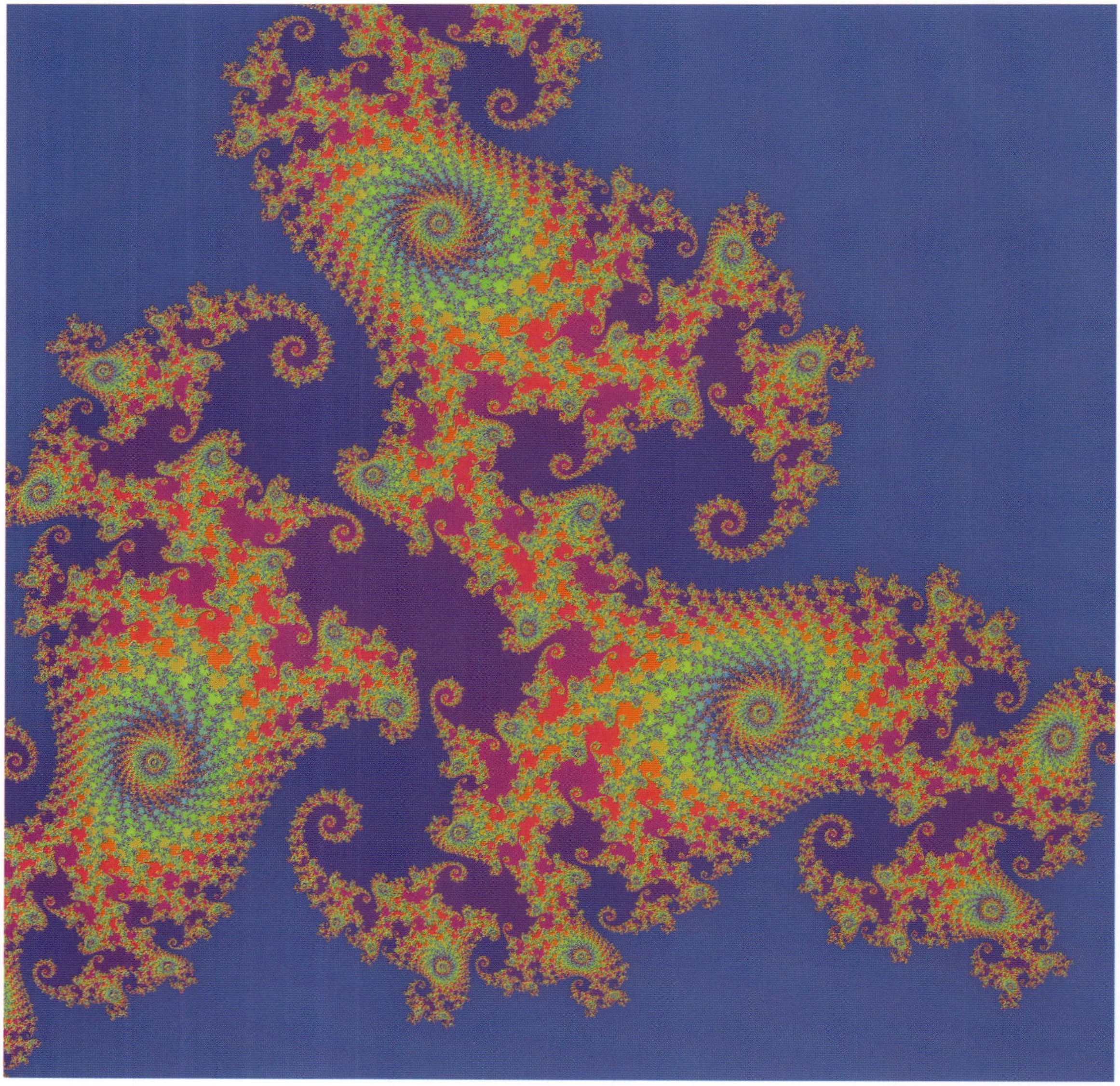

Skipping a Beat (Period-Doubling)

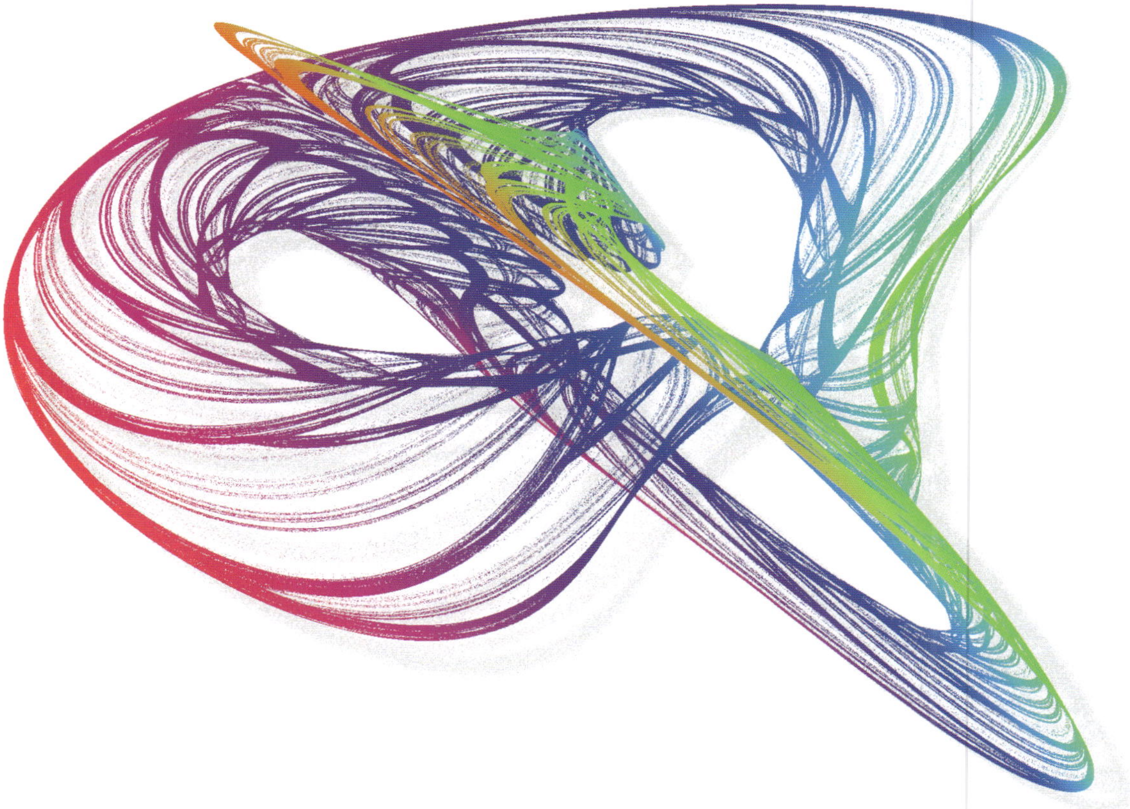

Period doubling

One of the most common routes to chaos is through a sequence of period-doubling bifurcations. A system that is periodic will suddenly change to a period twice as long (half the frequency) when a critical value of some control parameter is changed. A further change in the parameter causes the period to double again (to four times the original), and so forth, until the period is infinitely long, at which point the system is chaotic. A common example is a dripping faucet. At a slow drip rate, the drips are very regular like the ticks of a metronome. As you increase the flow, the drips occur faster, but at a critical value of the flow rate, the period doubles, and it sounds like "drip-drop, drip-drop ..." Further increase in the flow rate causes additional period doublings, until you eventually hear the sound of chaos where no regularity is evident. Try this experiment at home in your bathroom or kitchen.

THE ROUTE TO CHAOS

Period Doubling

Perhaps, like my uncle,
you have your ups and downs,
your good days and bad,
weeks of excitement
and weeks of dread,
years of enlightenment
and years of lost sleep —
even, like him, marry again
every few years.

Blue Sky Catastrophe

This is the day that dawns
just like any other day,
out to play with Honeybunch
and her new kittens,
run through the sprinkler
in noon's ninety-five degrees,
grass prickly under bare feet,
then in for dinner to find
your father's gone away
and won't be back.

Sensitive Dependence on Initial Conditions

You're eight, and no more
nightly games
of double solitaire with the dad
who's vanished now,
but because you can read
and the shelves are lined
with little red leather-bound classics
in 6 point type, you fall instead
into the worlds of Dickens and Poe,
Treasure Island and the Brothers Grimm,
the Swiss Family Robinson
shipwrecked on that island
where they all survive.

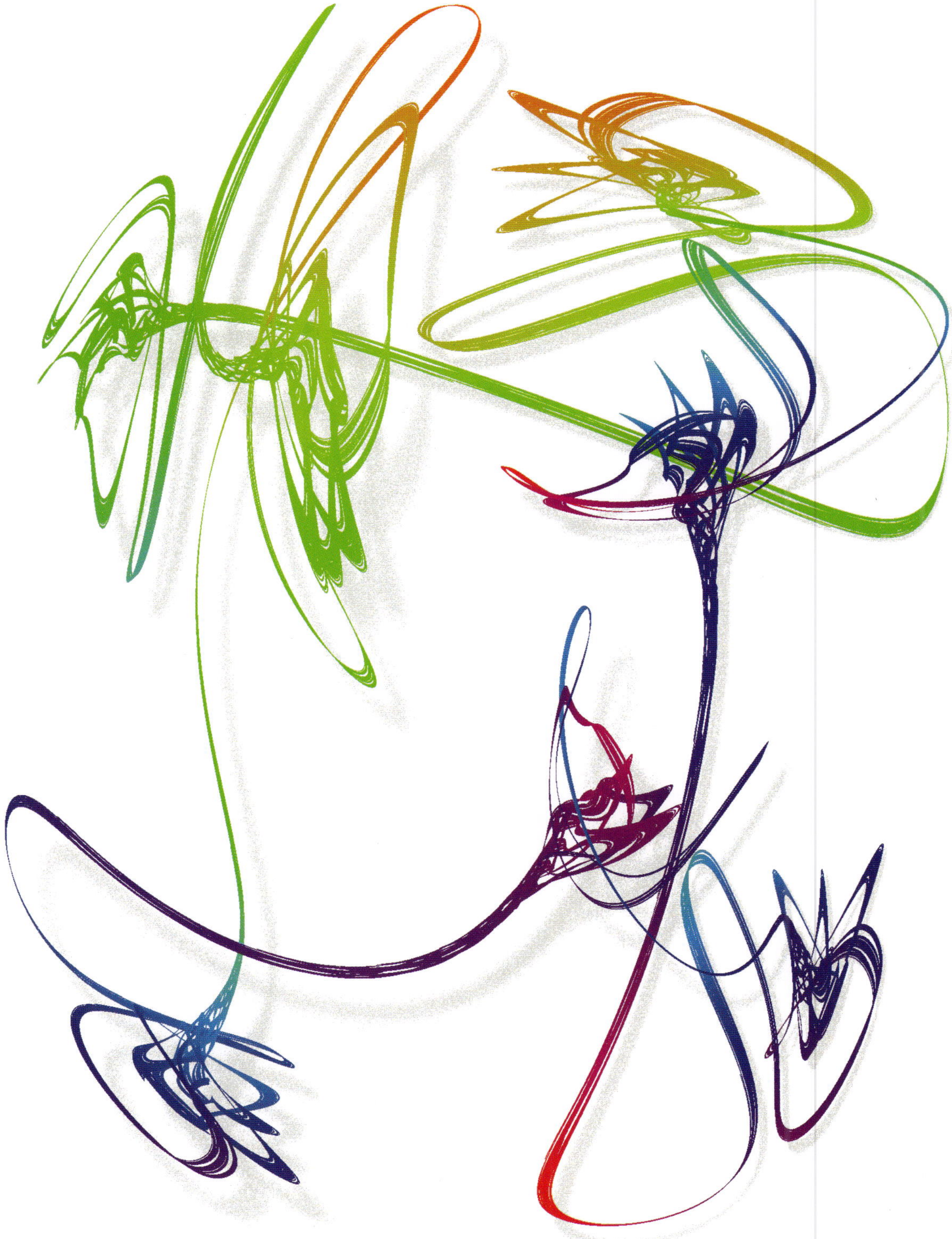

On the Edge of Chaos

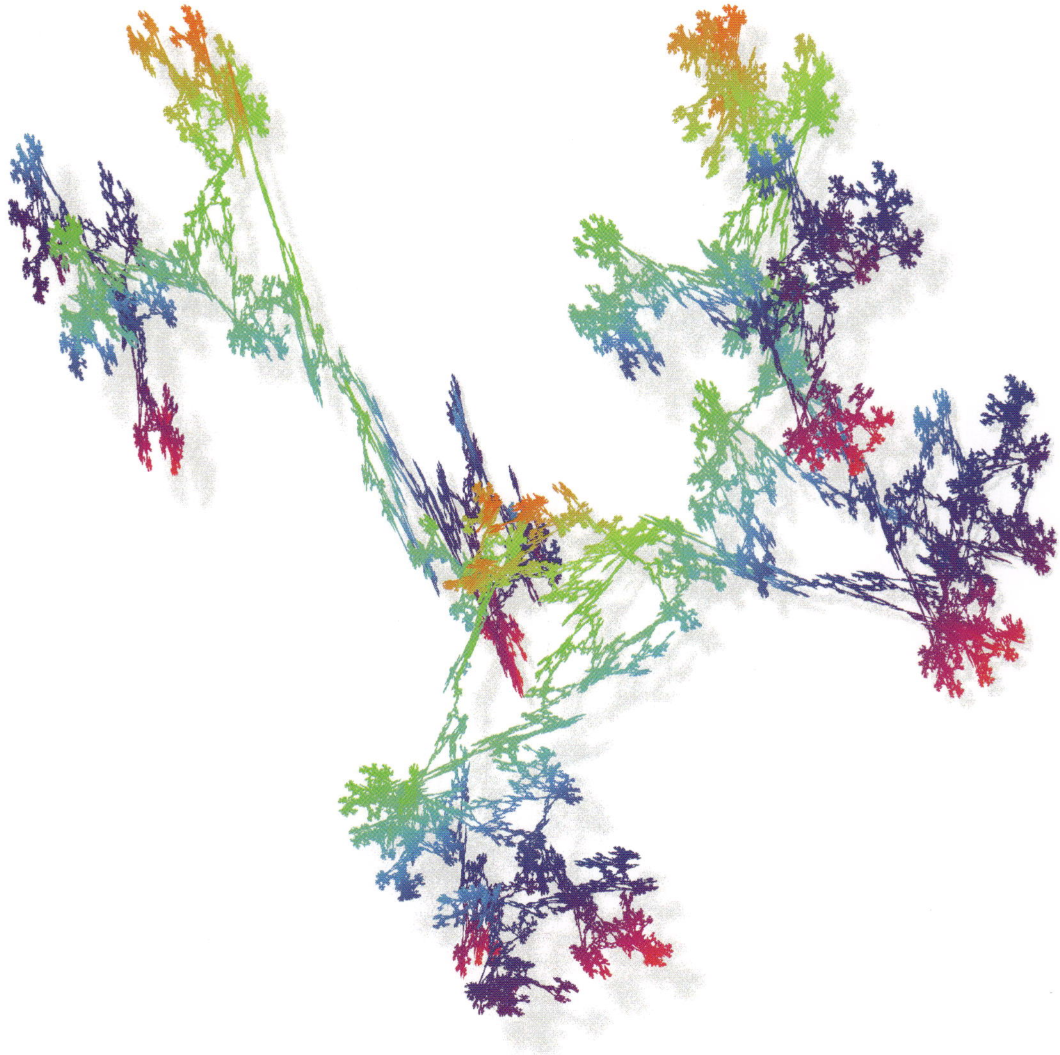

Self-organization

Self-organization occurs when an initially disordered (random) complex system begins to develop organized structures. For example, if you had a cellular automaton model of a forest landscape in which trees are initially scattered randomly in space, the model might develop spatial regions of dense forests interspersed with regions of prairies. It is a process that occurs for a system far from equilibrium in which there is a large throughput of energy or some equivalent resource. The energy typically enters the system on one scale (the whole landscape is bathed uniformly by sunlight) and leaves on a different scale (trees die one-by-one in isolated regions).

SAND BARREN

Flat on your stomach for a day
You'd see catchfly sleeping, Venus
Looking in her glass, the slow
Opening and closing
Of earthstars, ladies' tresses
Nodding to the hog-nosed snake
Who slides past the deer's
Double indentations,
Loosening the grains of sand
On the hoofprint's lip. In dreams
You'll see the avalanche.

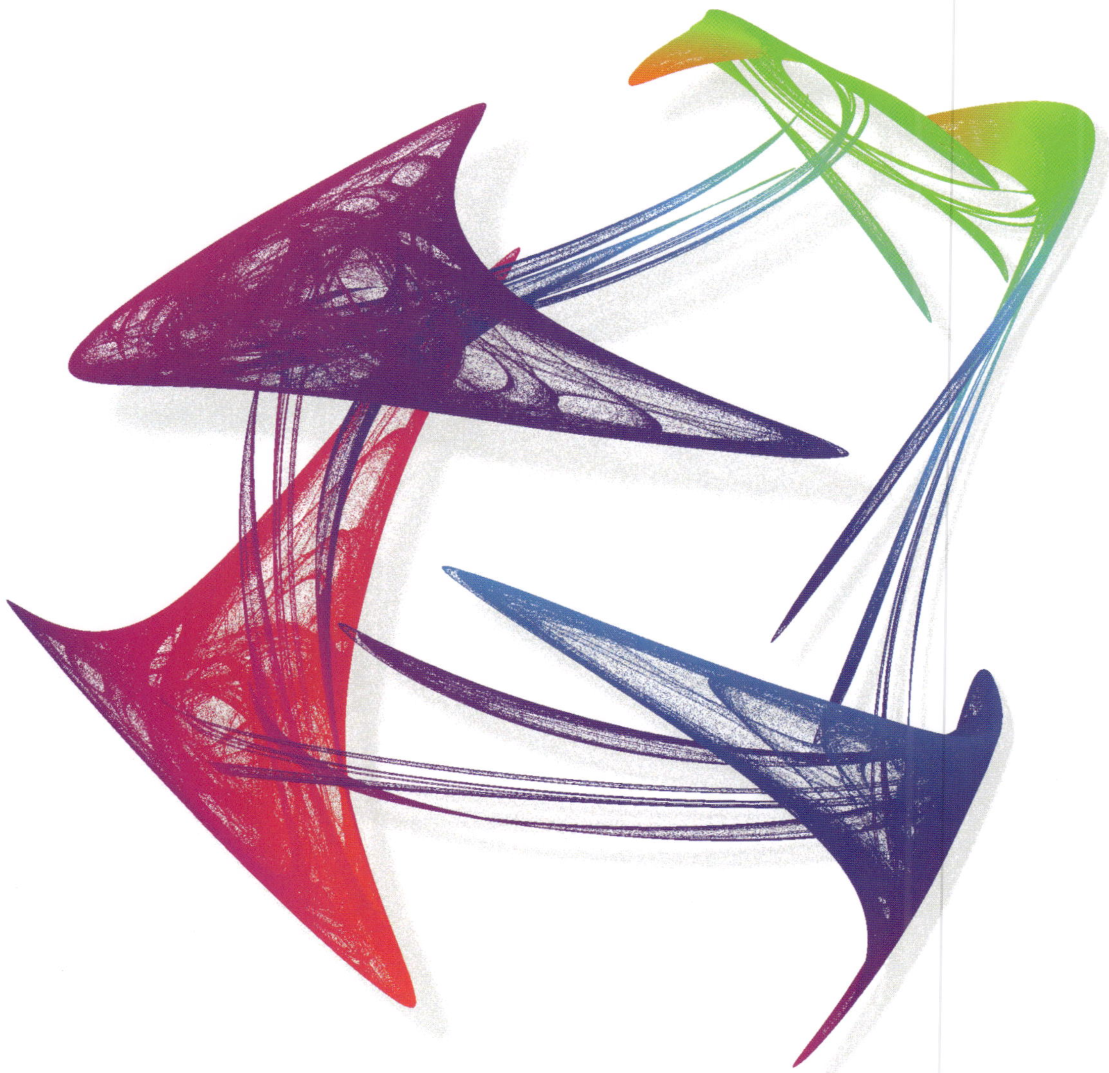

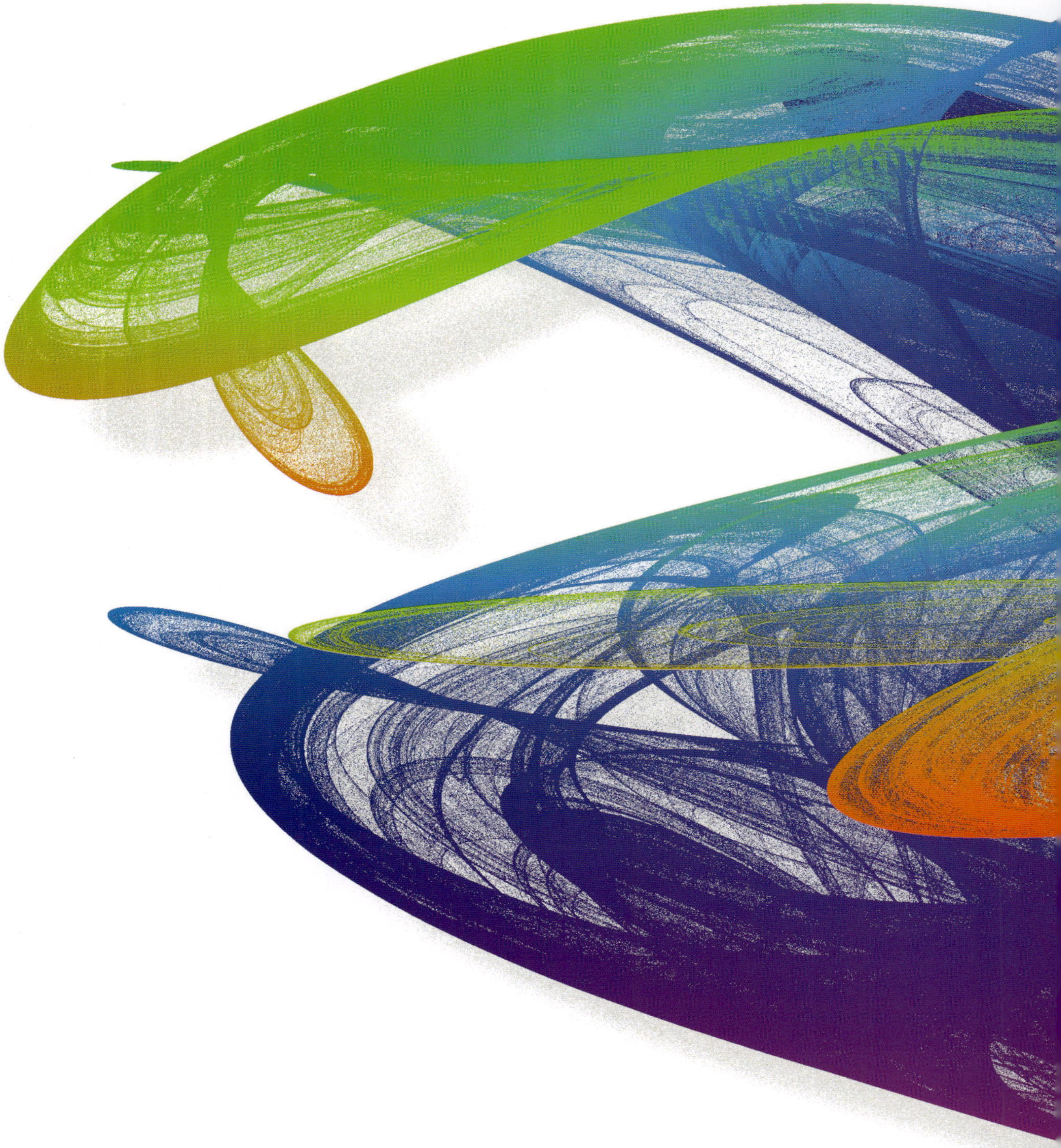

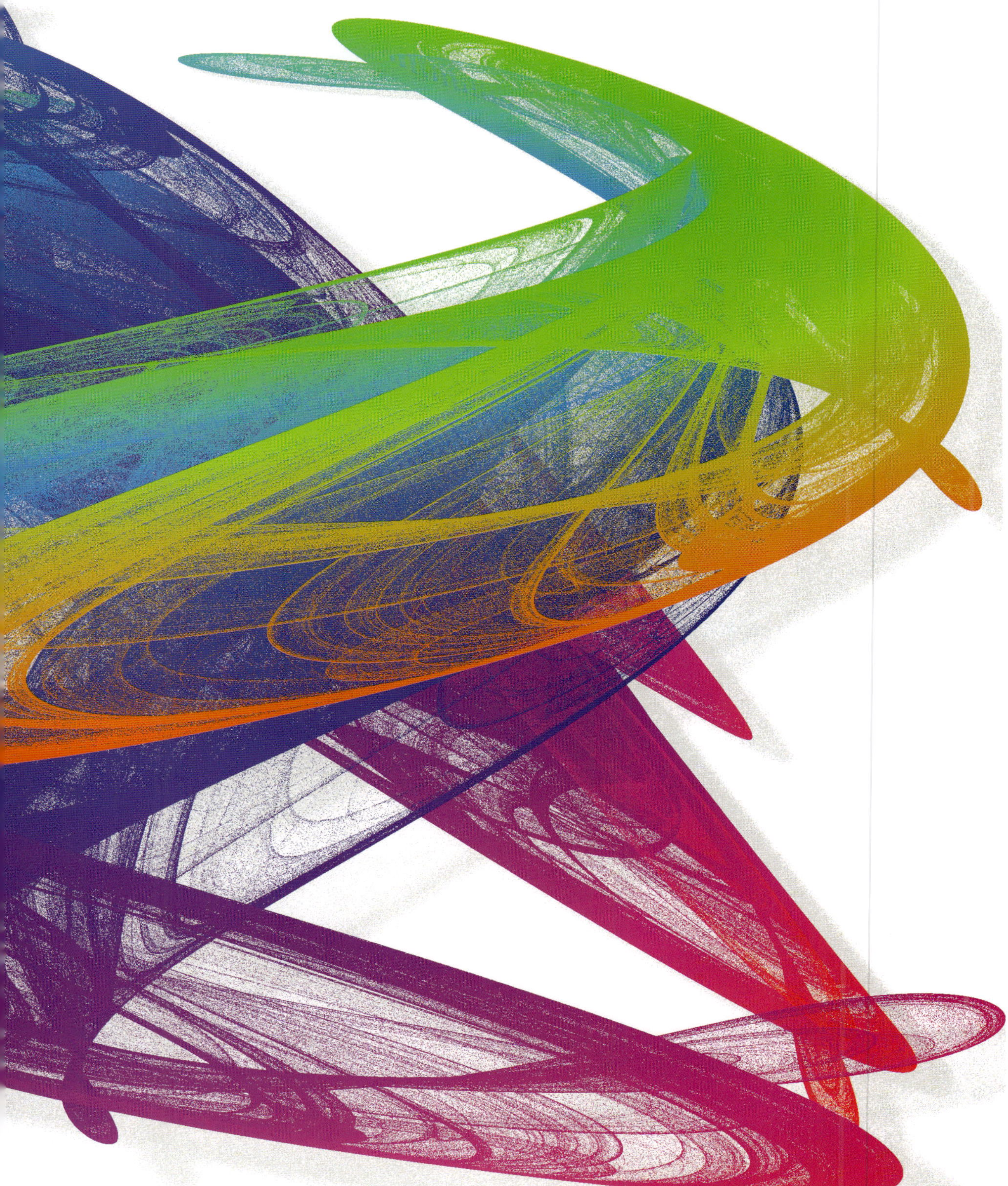

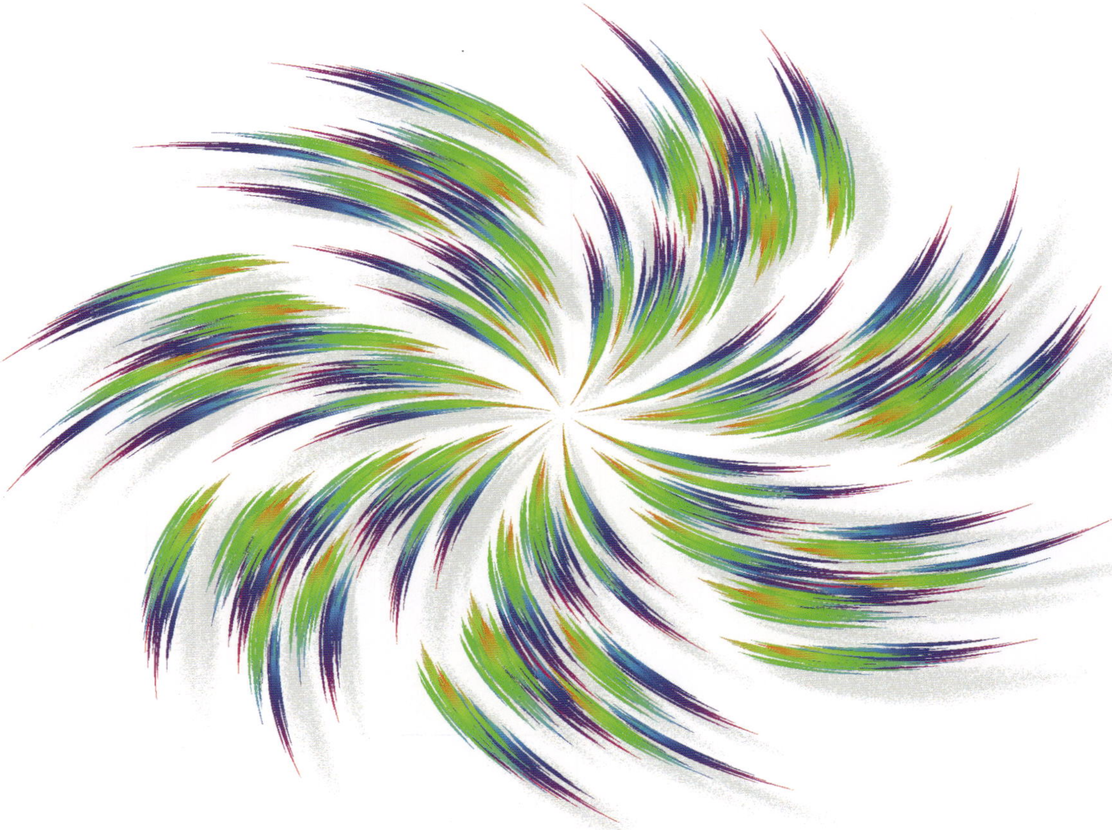

TREE

So you, too, balance uneasily
on the edge of chaos,
every cell computing the controlling field
of auxin's downward path from buds
defining 'up,' calling for new roots;
cytokinins' upward path from roots
demanding growth;
abscissic acid's report from leaf
of accumulated light, the turn
of season to winter's quiet —
and every cell adjusts
to whatever's reflected there
in the energy field —
beetles, shade, drought,
ice, the pruning wind,
the clearcut forest edge
with its shriveling microbial safety net,
sun and alluvial soil.

Self-organized criticality

A complex system that self-organizes typically does so in such a way that most quantities that describe it are scale-invariant, which means that they look the same on either small or large scales in size, time, and space (a fractal). The reason for this behavior is not well understood, but it seems to be a preferred state of nature. An example is earthquakes, for which there are many small ones and fewer large ones. In fact, the distribution of earthquake sizes obeys a power law (the number of earthquakes is inversely proportional to the size of the quake raised to some power), as is typical of self-organized criticality (SOC). Systems in such a SOC state resemble systems that are near a phase transition, with large fluctuations and strong connectivity.

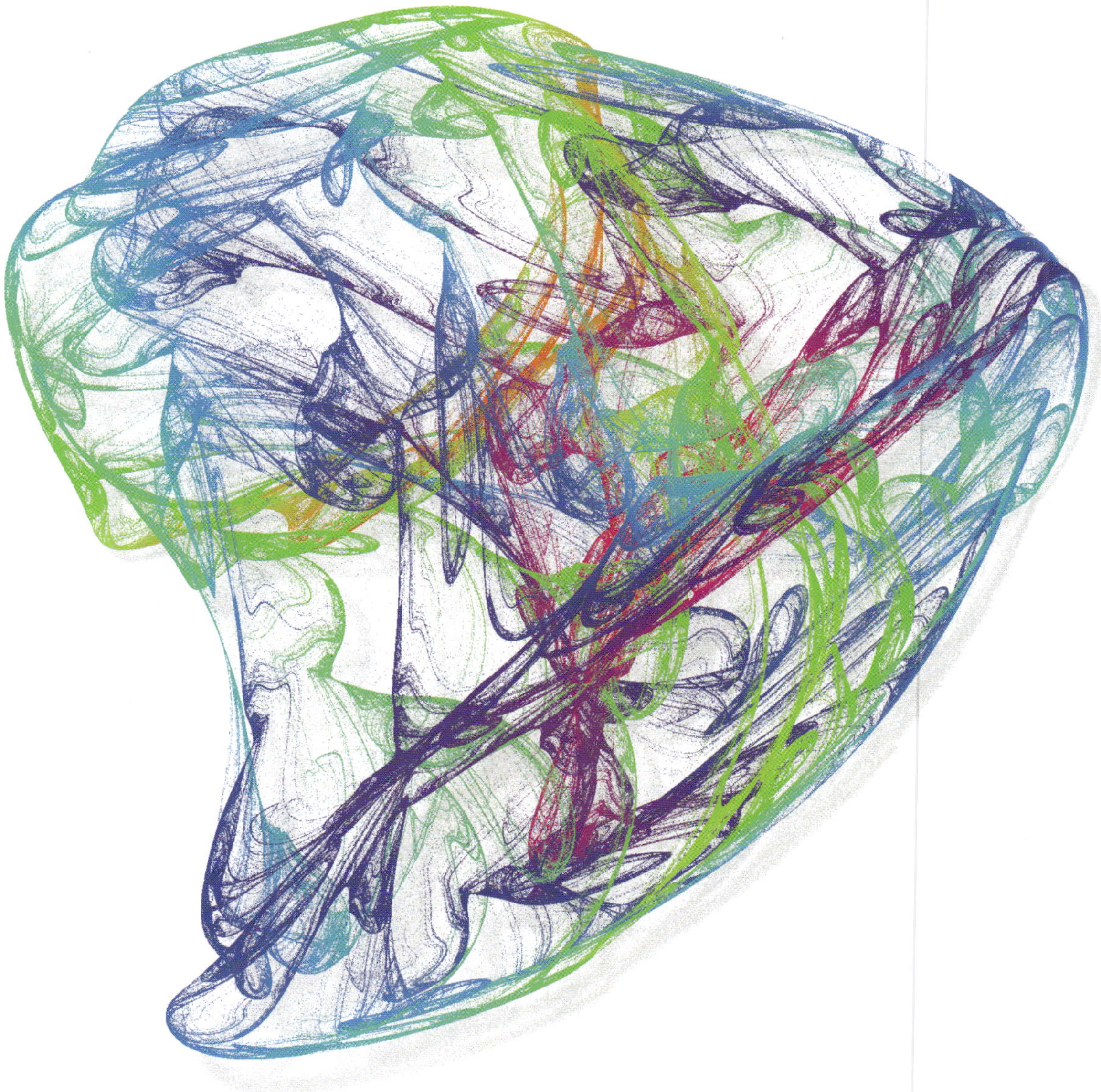

SELF-ORGANIZED CRITICALITY AMONG THE CHAOS THEORISTS

That verge of chaos where history
no longer predicts —
that's what Janine said, explaining to us
the evolution of forests in Wisconsin
while we finished our yoghurt and apples,
then cautiously asked each other, how
do you understand it? And Janine said
for her, SOC meant a process by which
complex structure emerged
from a random, disordered initial state
through repeated action of simple rules —
a neighboring plot comes to resemble
the trees on its sides — and a power law
describes the distribution of forest patches
of every size.

And I said, for me
it was that place on the graph
of the largest cluster of connected buttons,
plotted as a function of trials on which
you kept picking up any two, tying
the random pair together with string —
that place where the graph climbed
from small tangled clots of buttons
to one large knotted cluster suddenly,
just like that (waving my hands).

And Clint frowned and said, for him,
it was the sandpile poised on that edge
of collapse when one more grain started
avalanches of every size — that brink.

But whether oaks or buttons or grains,
time had brought us along
through small repetitive acts
of gladness and talk that
altered irretrievably some local patch
of lodgepole pine or solitary life;
then settled back, at the edge, again,
of whatever might happen next — life
on the edge of change at every scale.

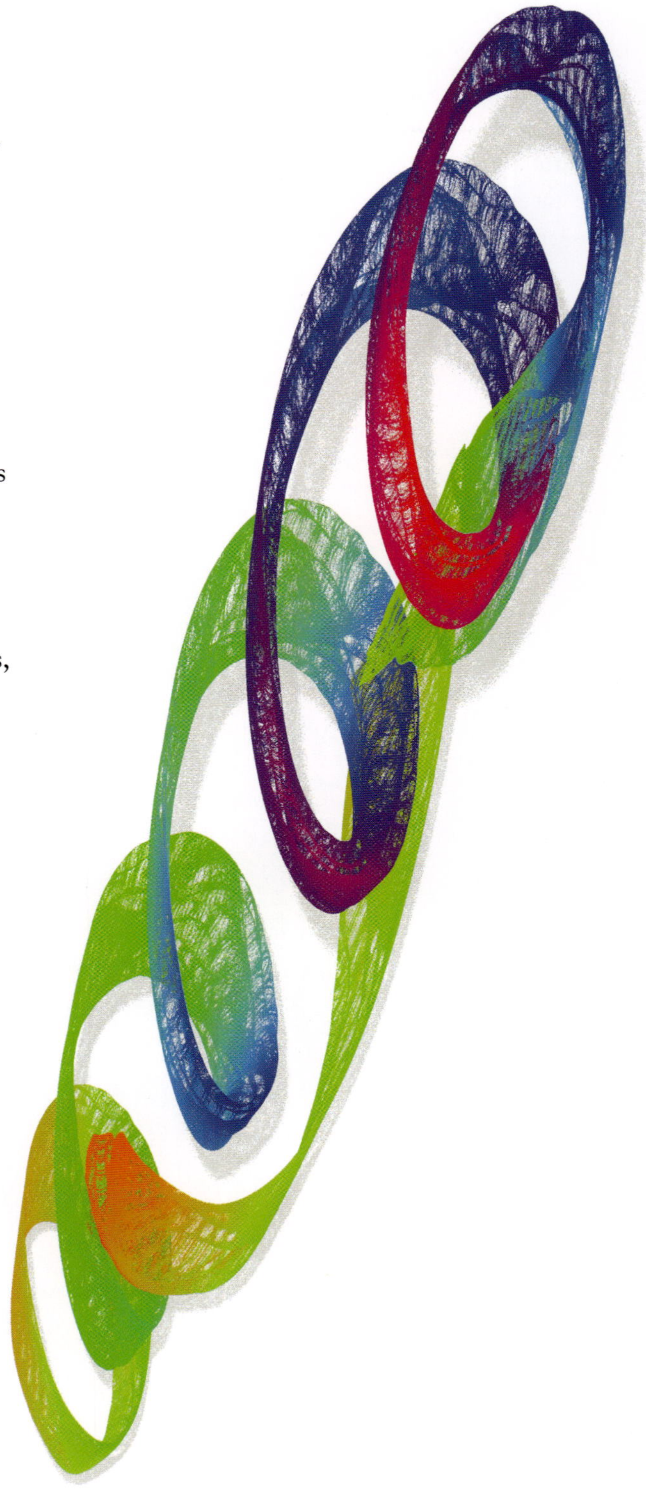

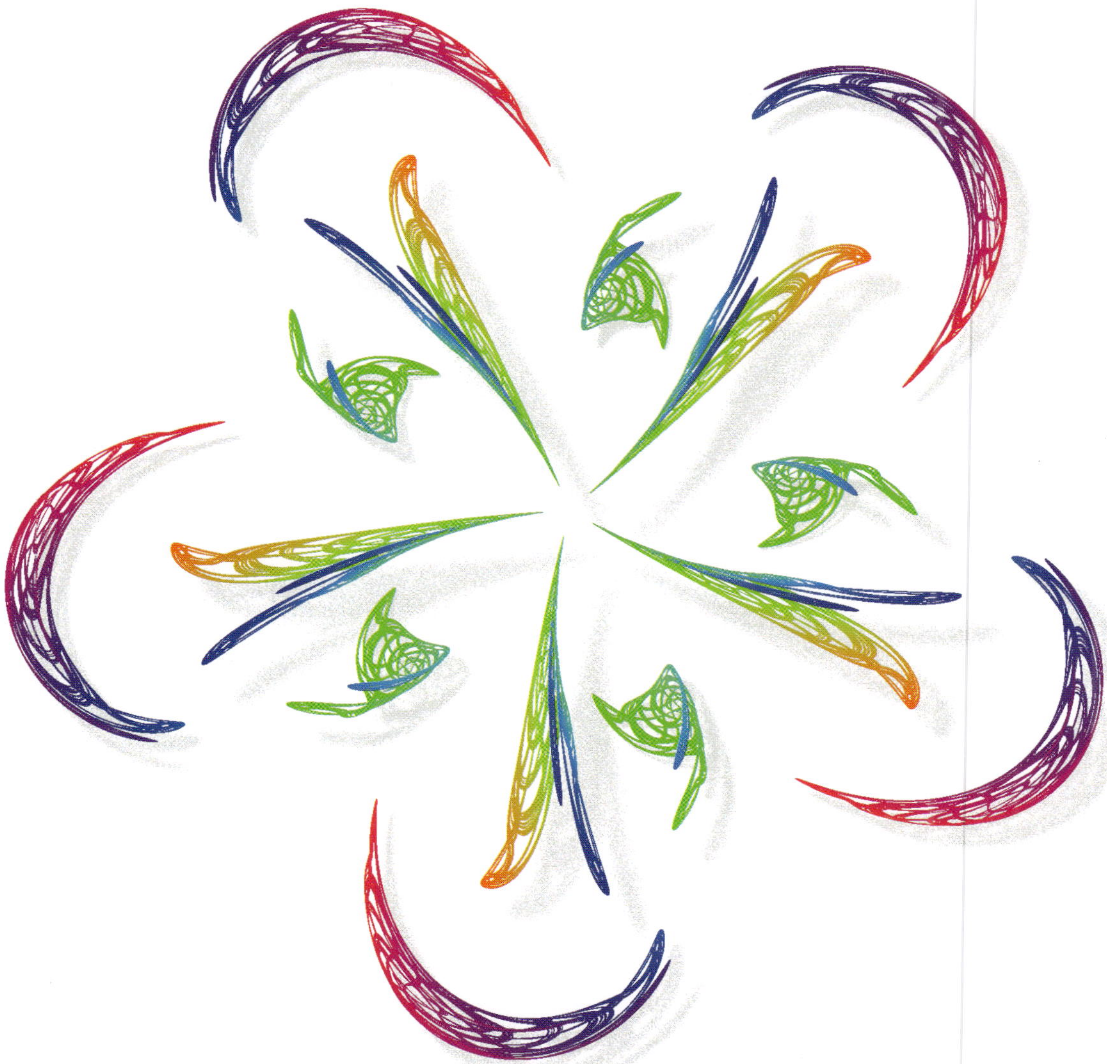

Chapter 5. Images of Chaos

The Universe is full of magical things, patiently waiting for our wits to grow sharper. — Eden Phillpotts, *A Shadow Passes*, 1934

Chaos or Noise?

Chaos

Chaos is a phenomenon that often occurs in deterministic nonlinear dynamical systems, in which a small change in the initial condition causes a very different outcome, as in the "butterfly effect." Errors in the initial conditions grow exponentially in time on average. Chaotic systems never repeat, and they are often sensitive to small changes in some parameter that controls the system. Chaos seems to be the rule in complex systems, but it also occurs in extremely simple systems with only a single variable if the system advances in distinct steps. When time advances smoothly and continuously, three variables are required for chaos. Chaotic systems are predictable for short times into the future, but not for long times. A good example of a chaotic system is the weather, for which predictions are almost worthless more than a week or so in advance. A simpler example is a dripping faucet if the drip rate is sufficiently rapid, in which case the interval between successive drips appears random and unpredictable.

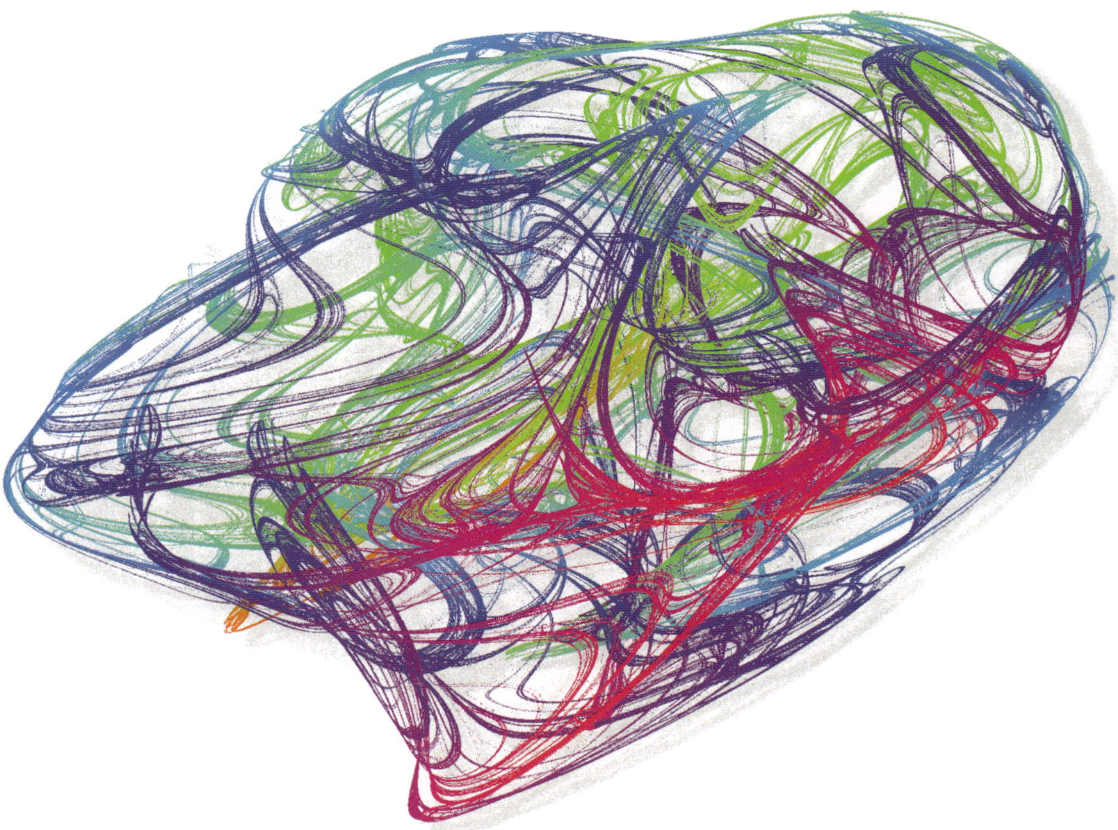

THE ART OF SIMULATED ANNEALING

> ***Def: a method in which***
> ***you explore shrinking Gaussian***
> ***neighborhoods of the best solution***

It's as if you took Coleridge's advice to heart,
"Best words in best order," and dropped the words
for your poem, *steel, heat, method, annealing,*
heart, tempered, into a jar,
shook them, vigorously at first,
then slowly, allowing each to find its place
in the rugged fitness landscape of Best Art
that could be made of these particular scraps
and dabs — though everything depends
on the Muse's steel-tempered method
and the heart's annealing heat.

Simulated annealing

When a liquid freezes, the atoms usually line up in an ordered array, forming a crystal with a minimum energy state for the atoms. If the cooling occurs too rapidly, the atoms do not have a chance to arrange themselves optimally, and the strength is compromised by defects in the crystal which is easily fractured. Consequently, metals are usually annealed by heating and then slow cooling to minimize such defects. A similar process occurs when you shake a sugar canister to increase the amount of sugar it can hold. The shaking is analogous to the thermal motion that occurs when a metal is heated. Simulated annealing is a computer technique in which a quantity is minimized by adding a random fluctuation to the parameters of the system and then slowly reducing the size of the fluctuation to optimize the solution. The method has been used to solve otherwise intractable problems such as the traveling salesman problem in which the task is to minimize the total distance traveled while visiting a group of cities.

THE TRAJECTORY OF A SINGLE PARTICLE

for Matt Briggs

It's all there since Newton's time,
Matt explains, gesturing —
the equations of motion that say
how the single particle, bumped,
will ricochet around the rest:

what every pool player knows,
leaning over his cue, calculating
angle and impact, but multiplied

so many times in a gas that
the smallest uncertainty of path,
that fraction of spin that sends
the next molecule left or right,
becomes, in split-moments,
doubled path, chaos, the random
walk of Brownian motion viewed
through the microscope. And thus,
Matt concludes, we can explain
the path of the particle exactly.
And haven't a clue,
this blink of time later, of what
will be the aftermath.

The pool player chalks a cue,
invites Matt to try his hand.

Determinism

A system is deterministic if its future is uniquely determined by its past. Sometimes determinism is contrasted with free will, and it is a deep and unanswered question whether biological and social systems are completely deterministic. Physical systems such as the Solar System are almost certainly deterministic, and the rules are very well known, allowing scientists to predict the motion of the planets and the occurrence of eclipses to high accuracy many years into the future. Of course one system can influence another, such as the climate influencing the ecology, and in such cases determinism occurs only on the largest scale, if at all. Highly complex systems may have millions of variables, and the determinism is often obscured by the abundance and complexity of the rules and interactions, making the behavior difficult to predict and apparently random.

Randomness

A system is random if its behavior is governed by pure chance. Randomness is the opposite of determinism. It may be that the only truly random processes are those that are governed by quantum mechanics such as the decay of radioactive nuclei. Other processes are effectively random, such as the flip of a coin or the roll of the dice. A random process may have some limited predictability. For example, a drunkard's walk may involve a random and totally unpredictable turn at every lamp post, but you can predict that the drunkard is very close to his previous position at each time step. Another example is the weather. It may be that some higher being is flipping a coin to determine whether it will rain tomorrow, but the coin must be loaded since we can make reasonably good weather predictions a day or two in advance. A sufficiently high-dimensional chaotic process, or one in which the Lyapunov exponent is very large, may be indistinguishable from random. In fact, such systems are used to produce the pseudorandom numbers provided in most computer languages.

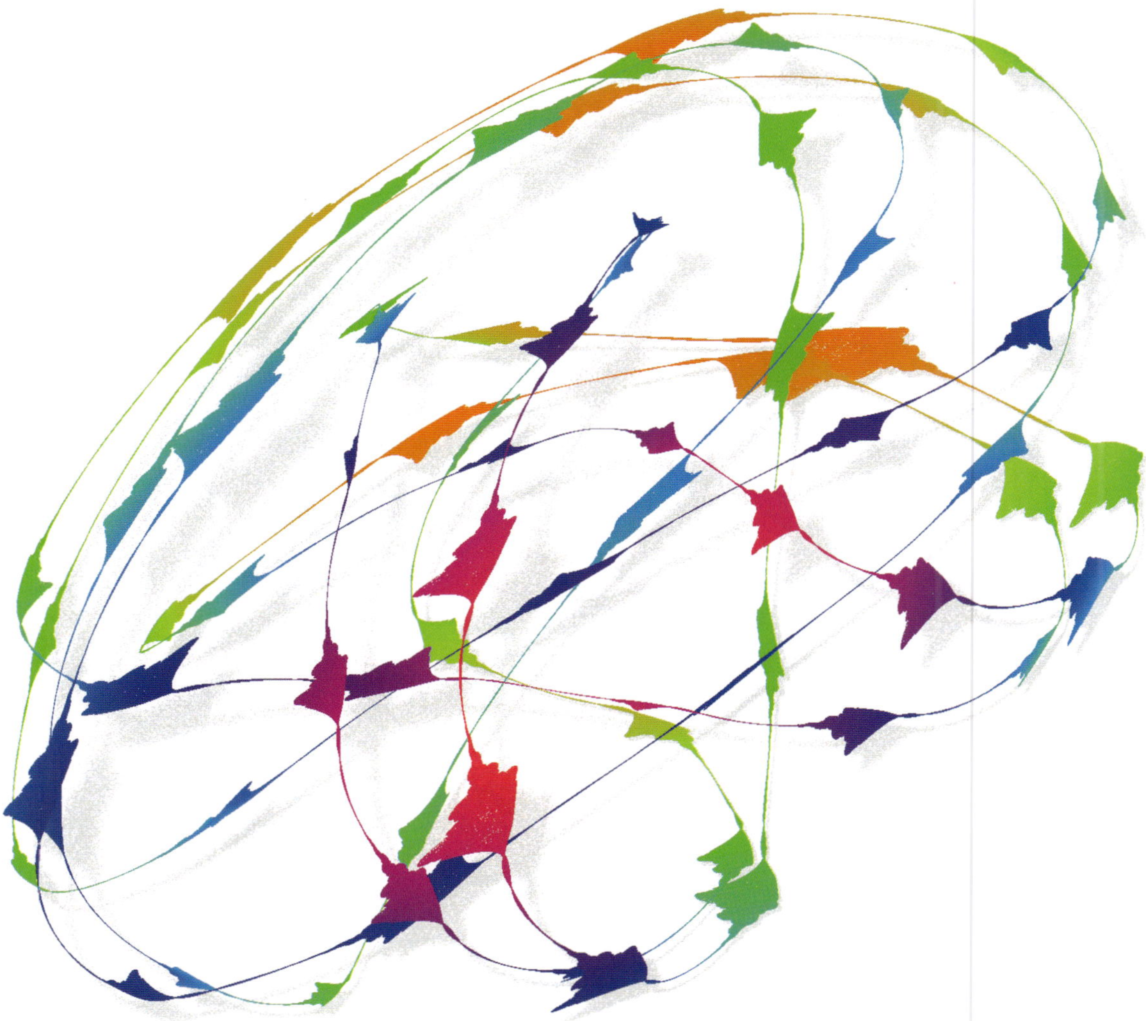

DIVIDING BY ZERO

1.

It was a childhood puzzle
before we knew
mathematicians called it
an 'illegal operation'
and banned its computation —
flashcards of stars in columns
and rows, blue and gold,
showing 8 or 4 divisible by 4,
by 2, by 1, or — zero?
Nothing goes into something
how many times?

2.

Let 'something'
be the all-in-all-of-it, the whole,
and get as close as you can
to nothing, now closer yet; divide;
the result blows up,
spilling over the edges
of the longest time and farthest space.
So surely zero into one
could create the universe —
all the stars and the human race.

3.

Or let 'something' be a lemon meringue pie,
brown-curled topping, tart filling,
tender crust — imagine it
waiting on some countertop,
owned by no one,
shared among none,
gathering dust — but no,
the mold will divide it equally
among its cells, if nothing else.

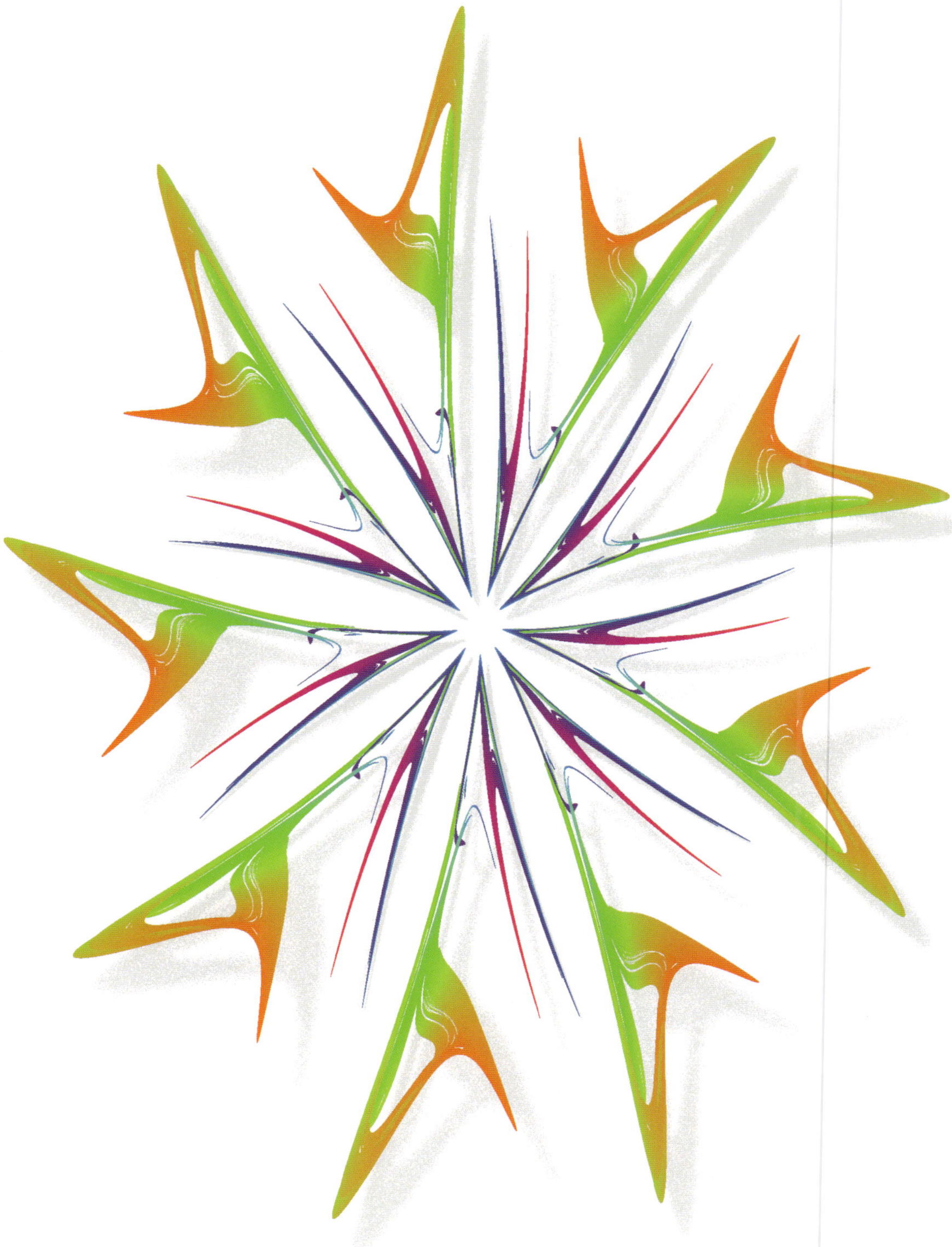

THE COMPOSERS DISCOVER CHAOS

Gone is the linear structure of time,
the old melodies, our rhythms and songs.

Composition becomes the moment's texture,
every performance a new work.

There are no more composers,
and musical notation has no power

to record chaos, though it points the way. The violinists,
almost unison, will close their ears and take us.

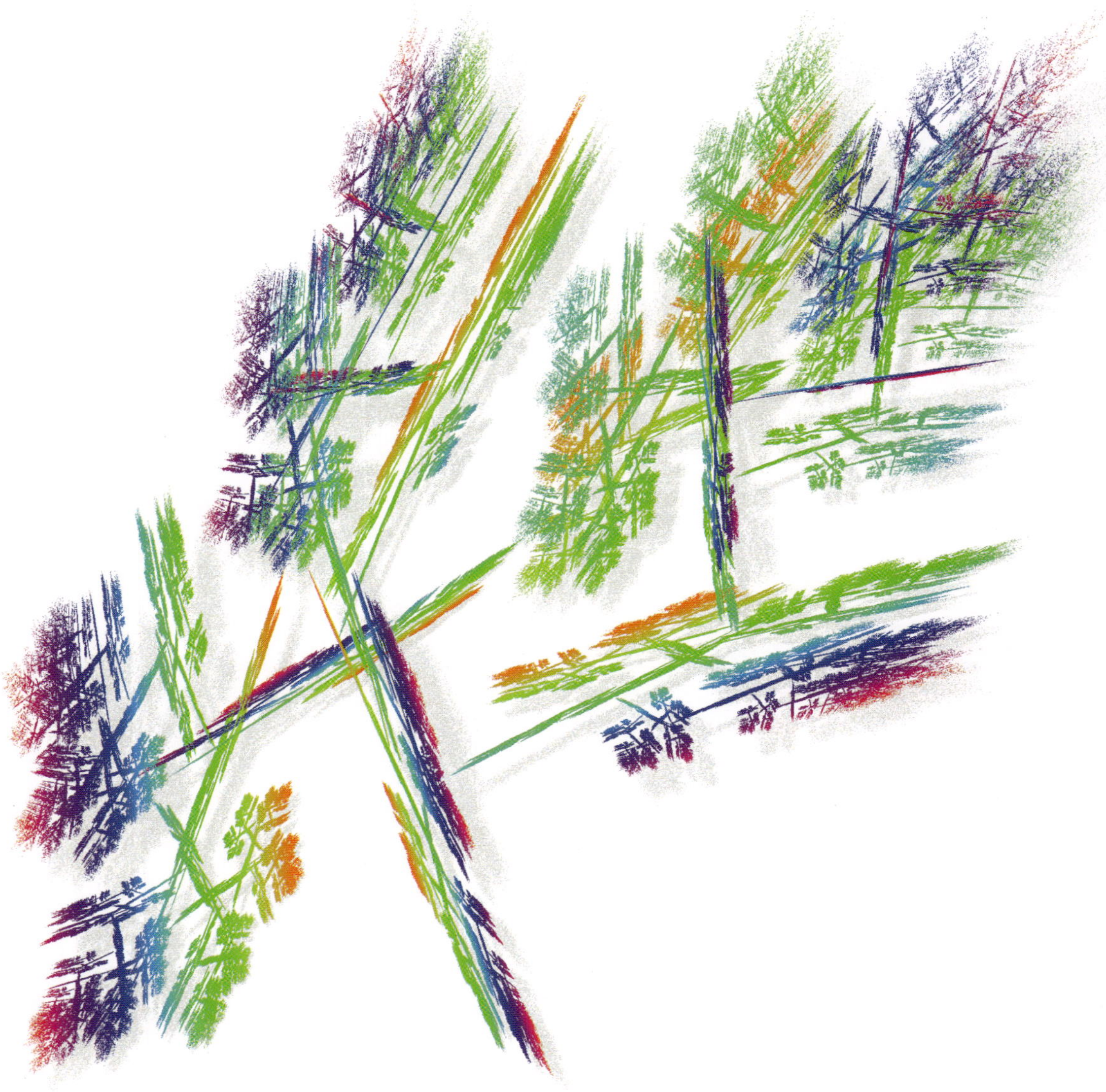

Sonification

Sometimes periodicity and chaos can be more easily detected by the ear than by the eye. Music and mathematics have a close connection since both involve rules and quantification. Brown music, named after the Scottish biologist Robert Brown (1773–1858) who discovered Brownian motion, consists of notes that change only slightly and randomly from the preceding note, much like an ant slowly walking across the musical score. White music, named by analogy with white light that contains all the colors of the rainbow, is what would be produced if all the notes were dropped into a hat and drawn out and played at random. Pink music is intermediate between the two. Most people find pink music most appealing, and composers seem to instinctively and unwittingly write their music in this form. Music can also be produced from chaos, which follows definite and precise rules, but that is unpredictable more than a few notes ahead. Periodic music is much too regular and predictable to be appealing. As with art, musical tastes tend toward a degree of order and structure, but with an element of surprise. Listen to the various types of music at http://sprott.physics.wisc.edu/chaostsa/fig09-07.wav and decide for yourself which is most appealing.

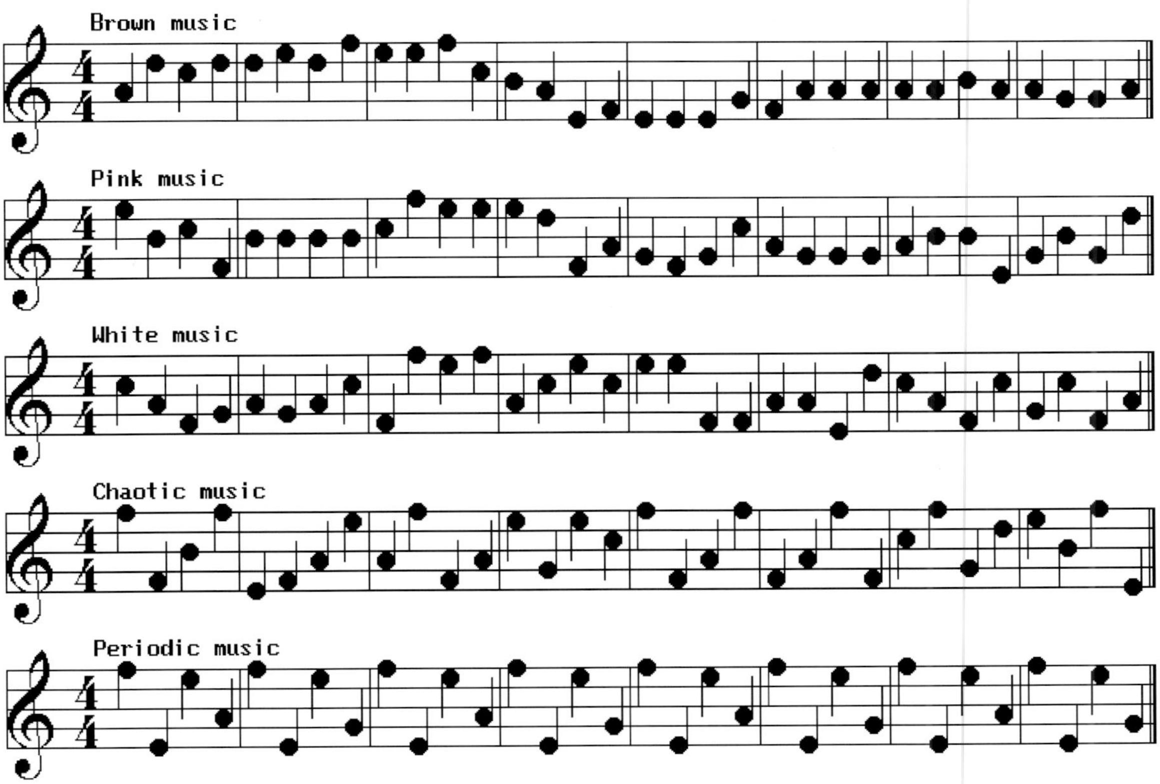

Mathematical music

Strange Attractors

Lyapunov exponent

In a chaotic system, small changes in the initial conditions make larger changes in the subsequent behavior that grow exponentially in time. The growth rate is called the Lyapunov exponent, named after Aleksandr Lyapunov (1857–1918). A deterministic system is chaotic if its Lyapunov exponent is positive, and the value of the exponent quantifies how chaotic the system is. The inverse of the exponent is roughly the time over which prediction is possible. There are actually as many Lyapunov exponents as there are variables (or dimensions) in a dynamical system, but only the largest (most positive) one is important for determining chaos, and it determines the rate at which points on the attractor separate as time advances. The negative exponents quantify the rate at which initial conditions are drawn to the attractor from different directions.

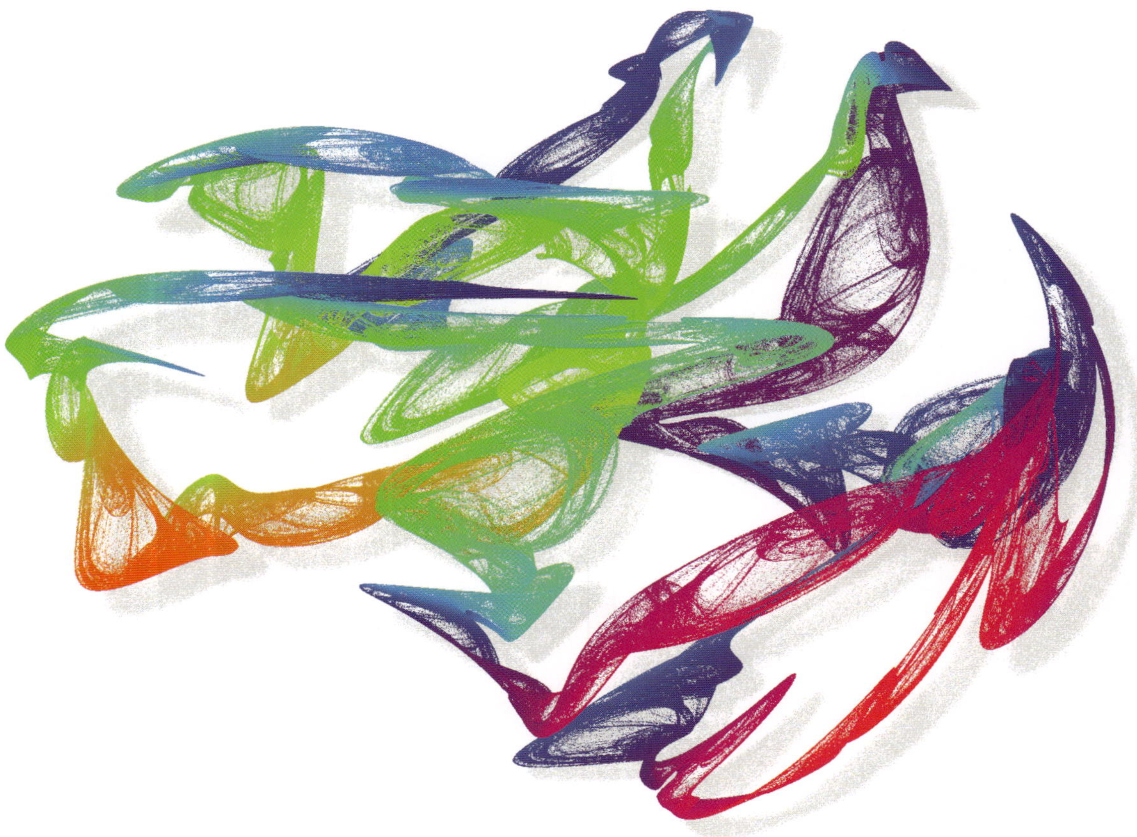

STRANGE ATTRACTORS

How to find them, those regions
Of space where the equation traces
Over and over a kind of path,
Like the moth that batters its way
Back toward the light
Or, hearing the high cry of the bat,
Folds its wings in a rolling dive?

And ourselves, fluttering toward and away
In a pattern that, given enough
Dimensions and point-of-view,
Anyone living there could plainly see —
Dance and story, advance, retreat,
A human chaos that some slight
Early difference altered irretrievably?

For one, the sound of her mother
Crying. For this other,
The hands that soothed
When he was sick. For a third,
The silence that collects
Around certain facts. And this one,
Sent to bed, longing for a nightlight.

Though we think this time to escape,
Holding a head up, nothing wrong,
Finding a way to beat the system,
Talking about anything else —
Travel, the weather — spending our time
At the flight simulator — for some
The journey circles back

To those strange, unpredictable attractors,
Secrets we can neither speak nor leave.

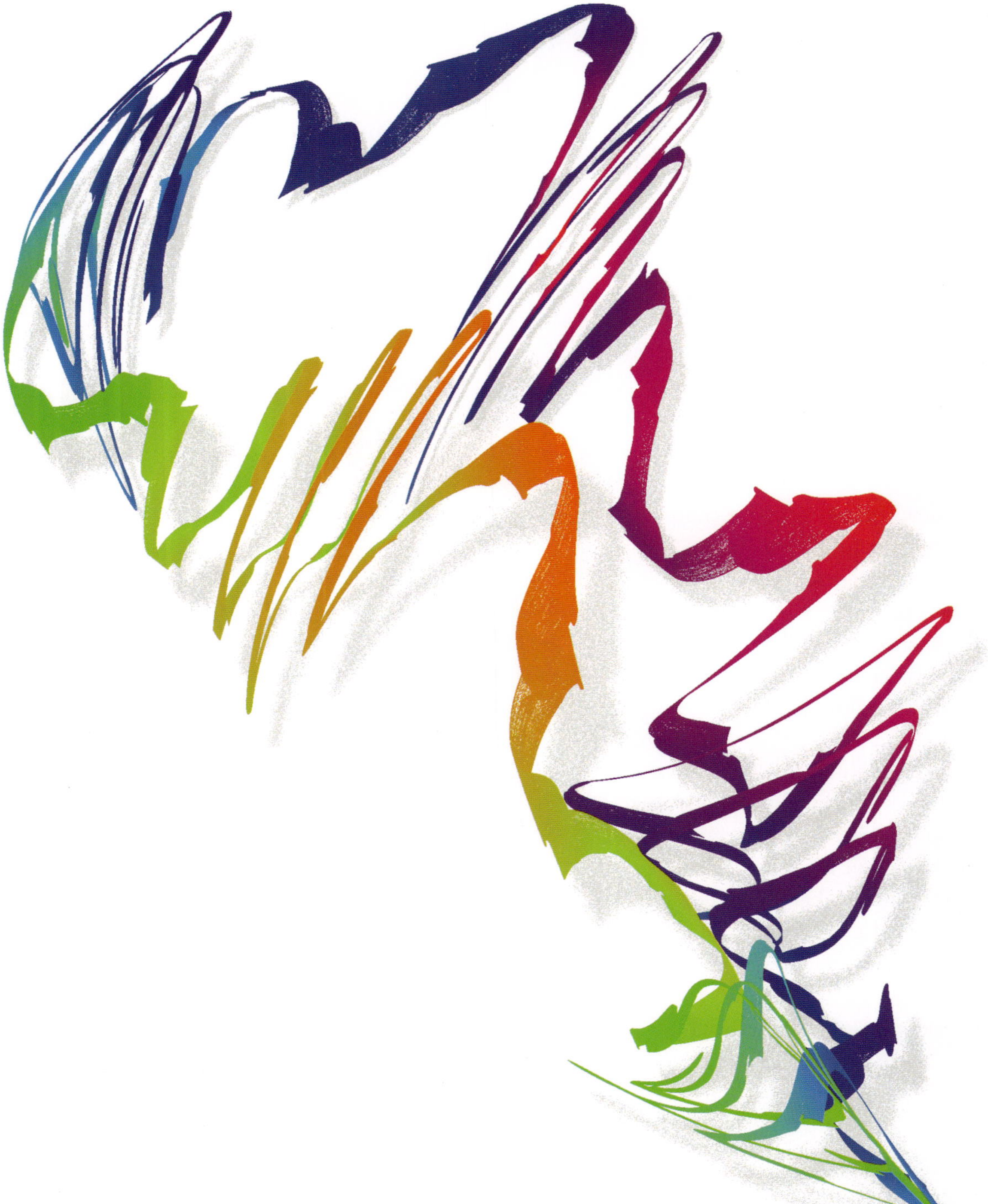

THE VOLATILE WIFE

$$X_{n+1} = X_n^2 - 0.2X_n - 0.9X_{n-1} + 0.6X_{n-2}$$

She is screaming into the phone at her spouse again,
anger raised to exponential height
because, in the first place, he'd said he'd be late,
damped now in memory,
and though, yes, he's called
to let her know he'll be later still,
there's the stirfry cooling on the table,
the rice growing sticky,
and looming intolerably,
the fact that he's later than usual —
though when he appears
she'll zap the plates in the microwave
and light the candles.

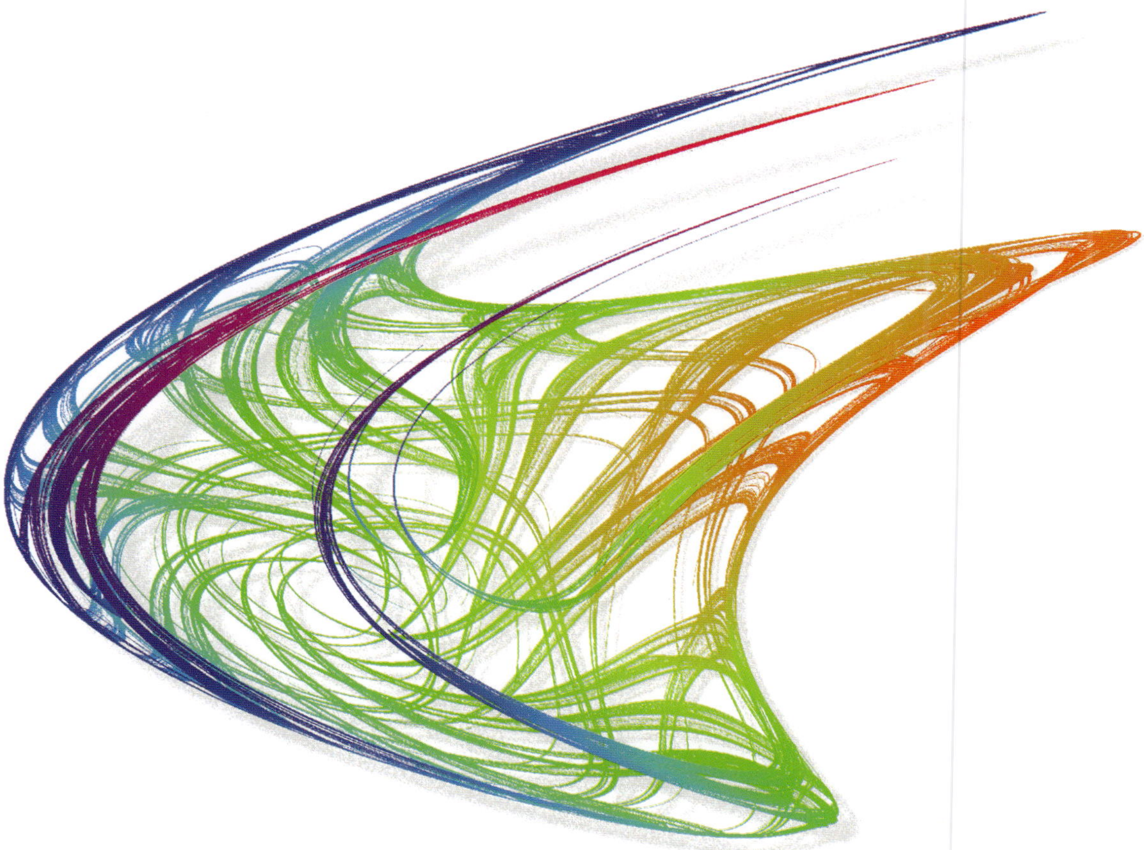

Strange attractor for the three-dimensional map
$X_{n+1} = X_n^2 - 0.2X_n - 0.9X_{n-1} + 0.6X_{n-2}$

Stretching and Moving (Iterated Function Systems)

THE NEAREST NEIGHBOR RULES

To your four nearest neighbors
offer keys, pet-sitting, lawn mowing,
snow-shoveling, house-watching
on vacation days, extra bloodroot.

To your eight nearest neighbors,
a hand raised in greeting,
praise for their flower beds,
inquiries about grown children.

To your sixteen nearest neighbors,
a wave on walks, comments
on their child's new bike,
visits in the spring and fall.

To your thirty-two nearest neighbors,
common cause in traffic signals,
nods in passing, park clean-up,
firewood scavenging.

To your sixty-four nearest neighbors,
a look in lit windows
on winter night walks, a knock
to offer voter registration.

Iterated function system

Imagine that you are located somewhere on a flat plane and you are given a set of rules that tell you where to go next from your current position. However, the rules are contradictory, so that you can only obey one at a time. Depending on the sequence in which you apply the rules, your route will be different, but if you consider all possible sequences of the rules (the iterated function system or IFS), you will end up visiting the same locations but in a different order. The resulting set of locations to which you are drawn from an arbitrary starting position is the attractor of the IFS, and it is usually a fractal with exactly self-similar structure on all scales. Many of the images in this book are from iterated function systems in which the rules are applied in a random order but repeated many times so that all sequences are eventually obtained.

Cellular automaton

One of the simplest models of a complex dynamical system is a cellular automaton (CA) in which a regular array of cells (like the squares on a checkerboard), each characterized by a state (such as black, red, or empty) that is updated at each time step according to some rule that usually involves the near neighborhood of the cell. The simplest CA is binary, with each cell having only two possible states. The rules can be deterministic or random, and the cells can be updated all at once or one at a time. The array can have any dimension, but one-dimensional (like the DNA molecule) and two-dimensional (like the checkerboard) examples are most common. The rules can be totalistic (only the total number of neighbors in a given state within some neighborhood matter), or more complicated. The neighborhood can be of any size, including the entire array. The set of possible rules is astronomical, and the corresponding spatial and temporal dynamics are extremely varied, ranging from trivial (all cells approaching the same state) to highly complex (apparently random and unpredictable).

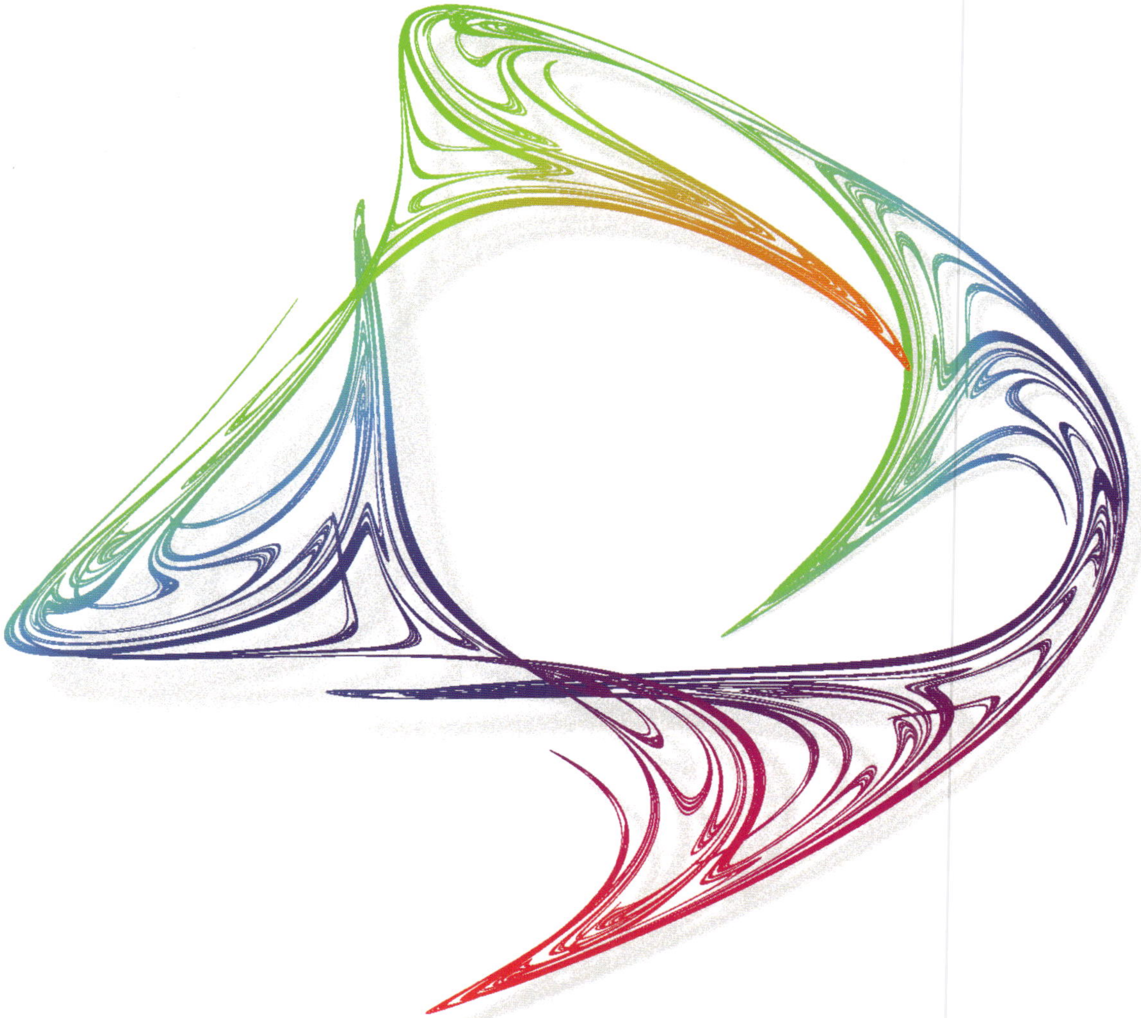

STOCKPILING AT Y-12

Imagine making ten thousand loaves
using robotic hands — that can,
at 30 yards, pour a beaker of water
into a flask without spilling a drop —
to gather the dough, to knead it
into shining metallic consistency,
slathering on layers of plutonium
like butter, rolling it thin as baklava,
sprinkling tritium like poppy seeds,
baking in the oven — loaves
to feed what hunger?

And then to stack them to cool
in secret cupboards
tended by remote control
as they age and mold,
decay to a half-life quarter-century
when, stale remnants, they
must be pulled out, dismantled
by those distant hands, rolled again
into the seeded stuff of death —
ten thousand thermonuclear loaves —

was it my best friend's father
who ran that national lab?
And my other best friend's father
whose hands ran the controls?
And me, a summer intern,
programming a computer there
to scale and plot data from any
countable objects — how did we not
know? Or the prudent townspeople
not speak of it? The work
that put bread into our mouths.

Escaping the Attractor (Generalized Julia Sets)

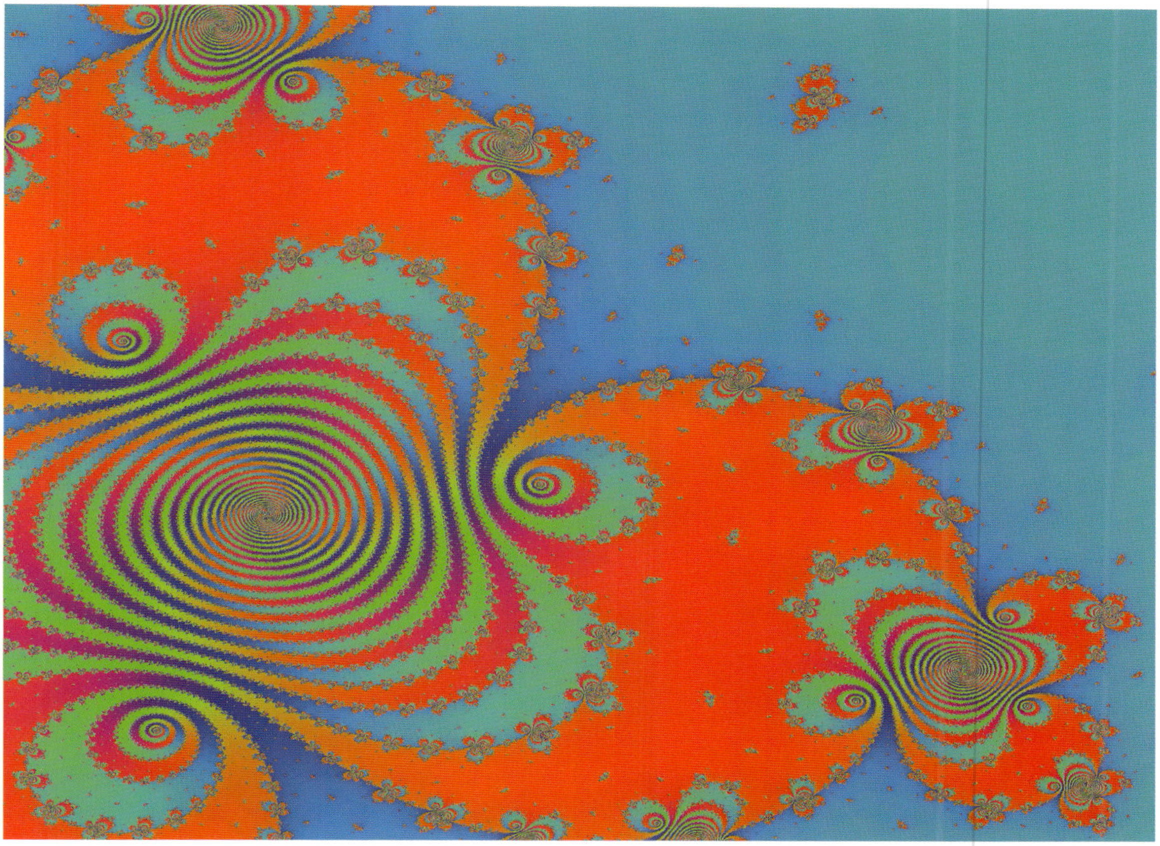

Julia set

A Julia set, named after Gaston Julia (1893–1978), is the boundary of the basin of attraction for bounded solutions of a particular iterated function in the space of possible initial conditions. The usual function involves a complex variable (one with a real and imaginary part) and a complex parameter that controls the shape of the basin boundary. There are infinitely many Julia sets, one for each value of the complex parameter, which itself has a real and imaginary part. The Julia sets are usually fractals with an especially intricate and beautiful structure. Sometimes, the interior of the Julia set is plotted instead of the boundary, in which case it is called a "filled-in Julia set." More often, points are plotted outside the Julia set, color coded according to the number of iterations required for the orbit to cross some boundary, after which it is known or assumed to go to infinity. Such plots are called "escape-time plots," and many examples of them are shown in this book.

BASIN OF ATTRACTION

for Jon Foley

The lake is the attractor for the rain.
It waits as cloud freight loosens and falls.
It calls the raindrops down the hills,
Gathers acolytes in gutters and drains,
Seeps, streams, gullies, creeks,
Receives them all, all that have found
The divide that leads them along the ground
And into the waters of the lake.
Though the raindrops trace new trails,
The wind scours wave and soil to make
A changing watershed, ice rakes
The rock of borderlands. Fog veils
The roots and leaves that remember rain
And send it back to cloud and lake again.

Basin of attraction

Surrounding every attractor is a basin of attraction, which consists of those initial conditions that are drawn to the attractor and will eventually reach the attractor for all practical purposes. You can think of the basin as resembling the basin of a bathroom sink that collects into the drain all the water that falls into it, or the watershed of a lake. The basin may stretch to infinity in one or more directions, and the boundary of the basin (like the continental divide) is often a fractal. Although it is easiest to think of basins in two dimensions such as the surface of the Earth, they occur in systems of arbitrary dimension. Initial conditions outside the basin of attraction are drawn to a different attractor or perhaps to infinity. A dynamical system may have many coexisting attractors, each with its own basin of attraction (like the Atlantic and Pacific Ocean basins), and the boundaries can be intertwined in a very complicated way as with the Julia sets displayed in this book.

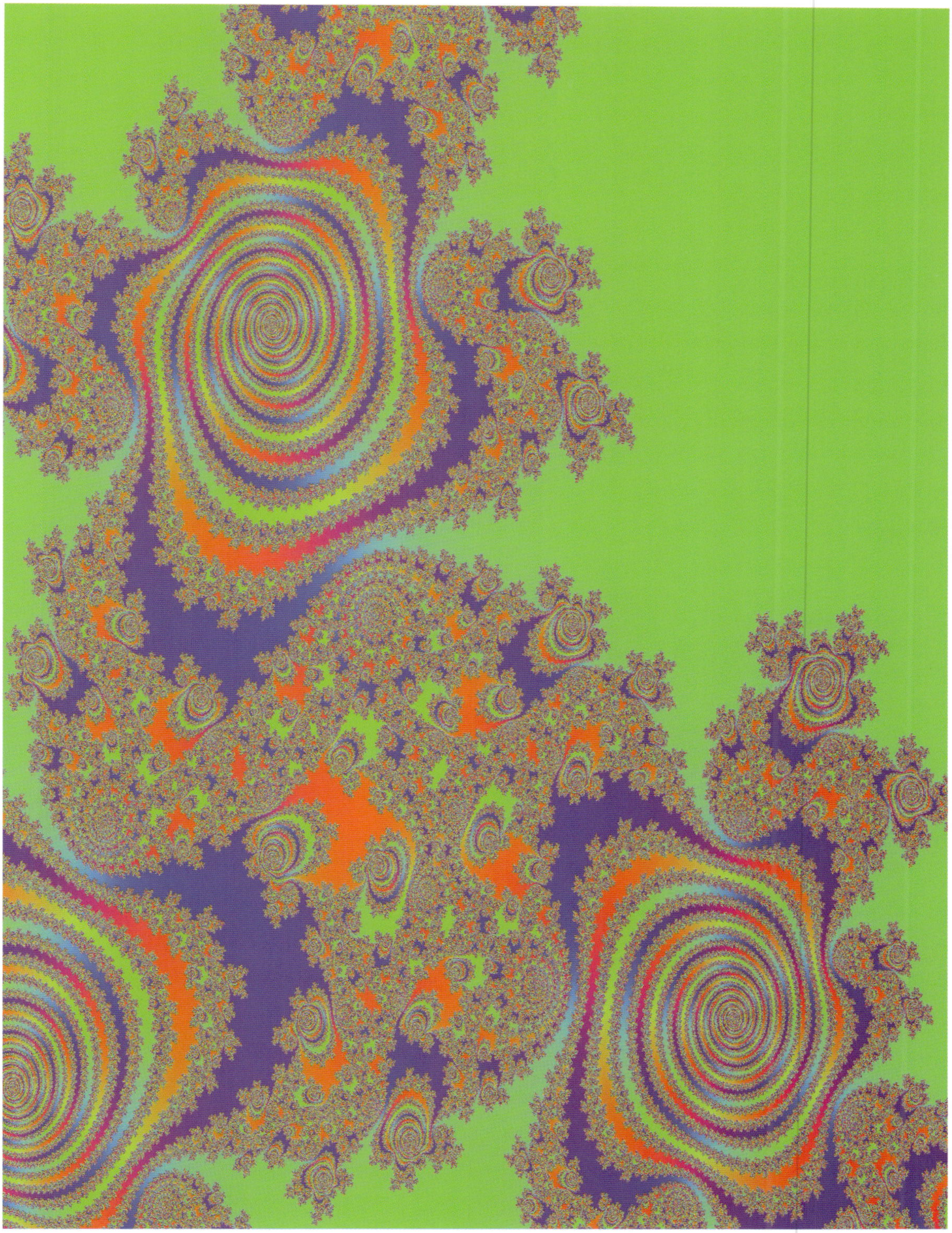

ESCAPING THE MANDELBROT SET

Def: Equation yielding a fractal pattern self-similar at every level of magnification within a range of values.

She says
The coffee is fine
Though it could have been stronger
And cream would be nice.

She says
The weather today
Is, yes, fine, though cold
For summer and more rain likely tonight.

She says
The summer's going well,
Of course awfully fast and won't last
Long enough to get done what she'd planned.

She says
The marriage was ten good years
And then ten bad, and she's learned
A lot since, though of course its lonely.

She says
Buying a new cappuccino maker,
Espresso roast, and best jam for her bread
Is frivolous, but we only have one life.

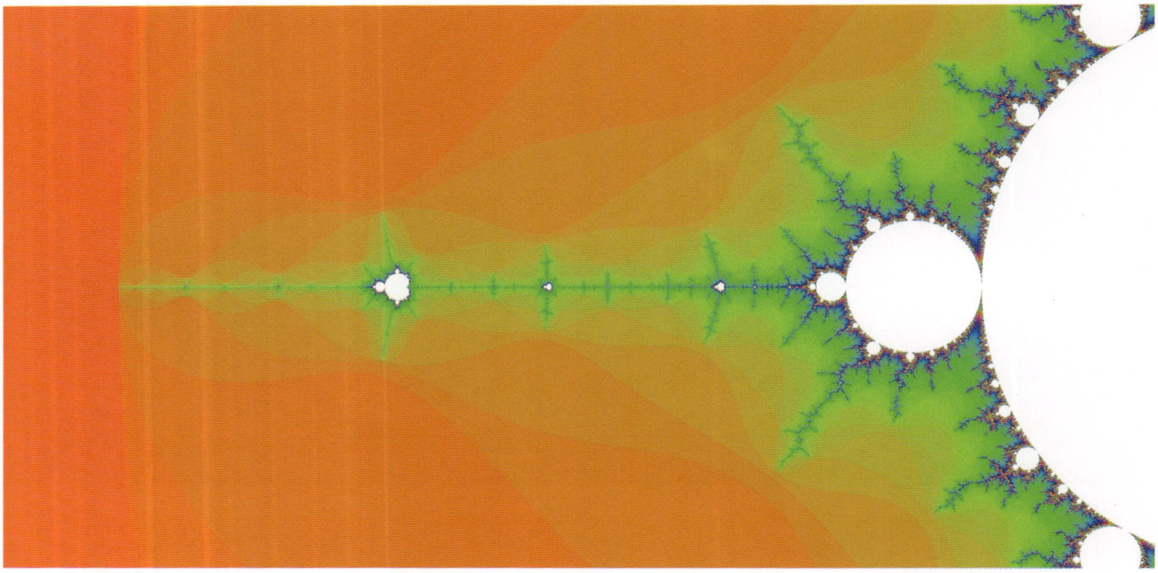

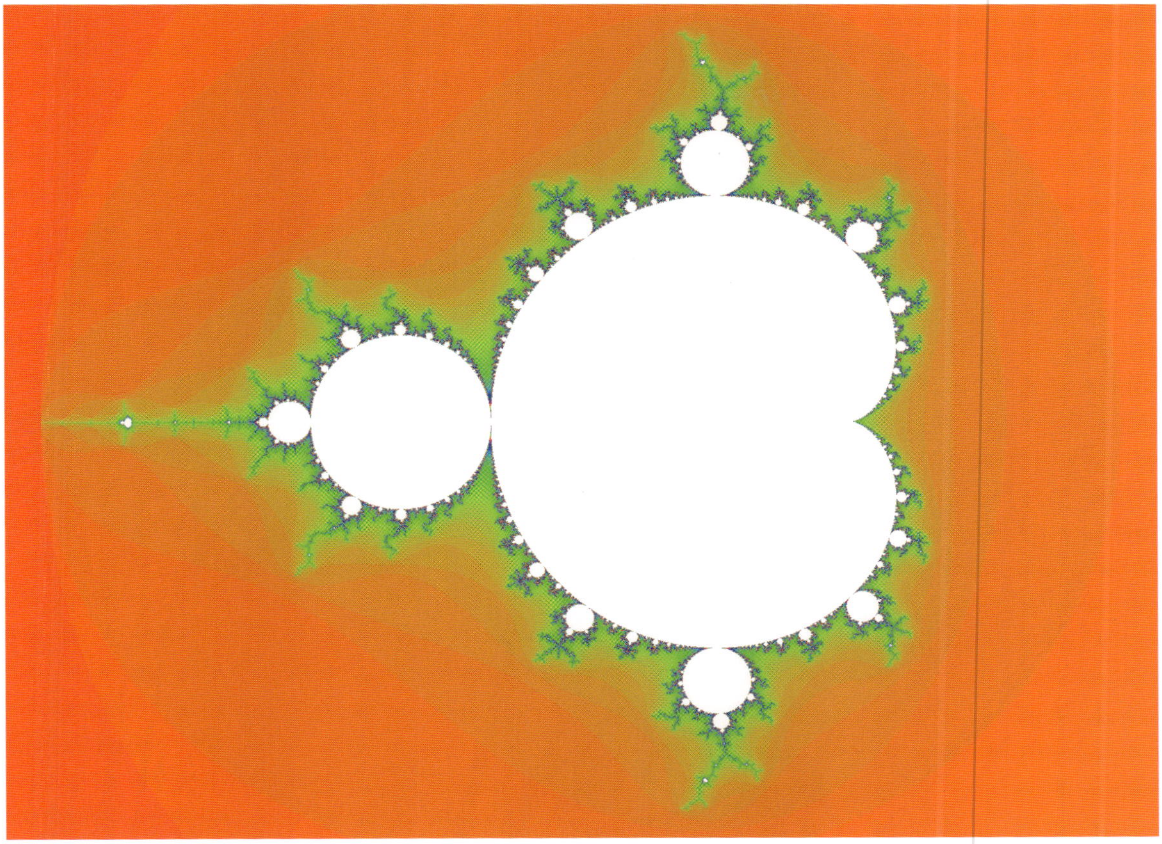

The Mandelbrot set

Mandelbrot set

Julia sets are of two types, those whose interiors are connected (one large region), and those whose interiors are disconnected (sometimes called "dusts"). The Mandelbrot set is the set of all parameters whose Julia sets are connected. Regions near the boundary of the Mandelbrot set correspond to those Julia sets that are just barely connected or barely disconnected, and those regions contain the most visually interesting fractal images. Hence the Mandelbrot set is a map of all the Julia sets. As with the Julia sets, the Mandelbrot set is usually exhibited with colors or gray levels indicating the number of iterations required for the orbit to cross some boundary, after which it goes to infinity. It is perhaps the most famous of all fractals, and zooms into the region near its boundary reveal extraordinarily intricate detail. In fact, as you zoom deeper into some region of the Mandelbrot set, the image increasingly resembles the Julia set corresponding to that region. In that sense, the Mandelbrot set contains all the Julia sets.

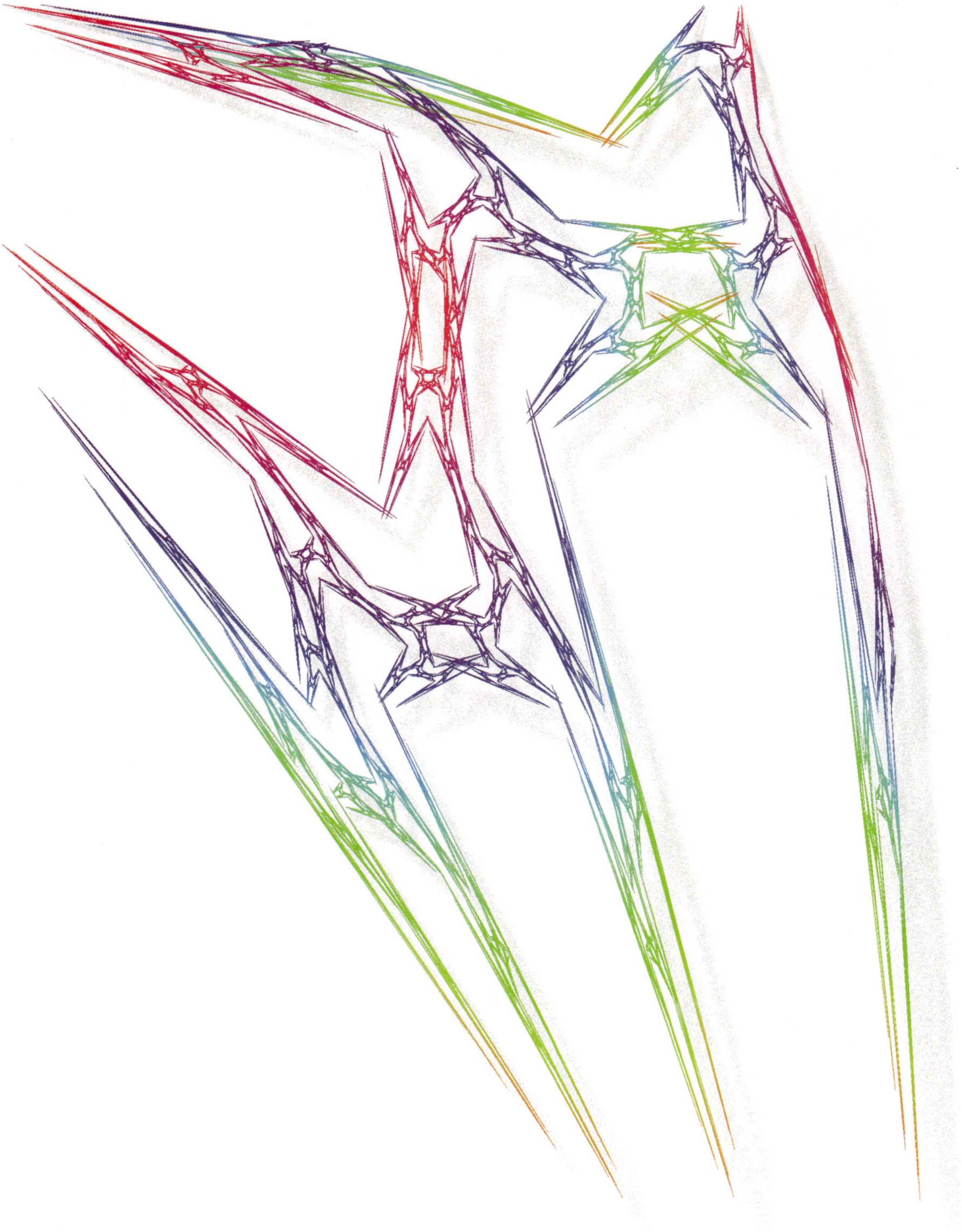

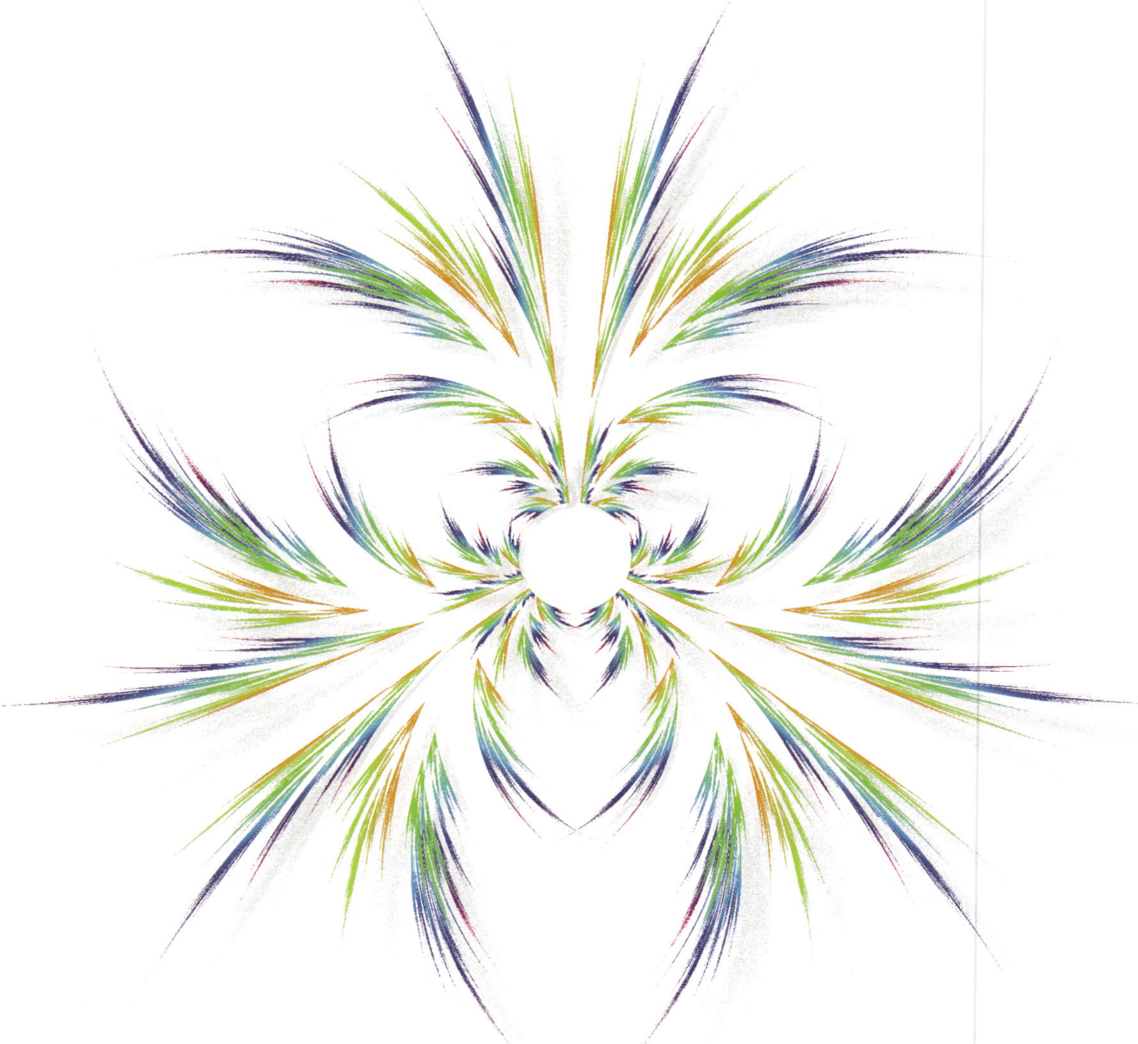

Chapter 6. Chaos and Predictability

In these matters the only certainty is that there is nothing certain.
— Pliny the Elder

Time's Arrow

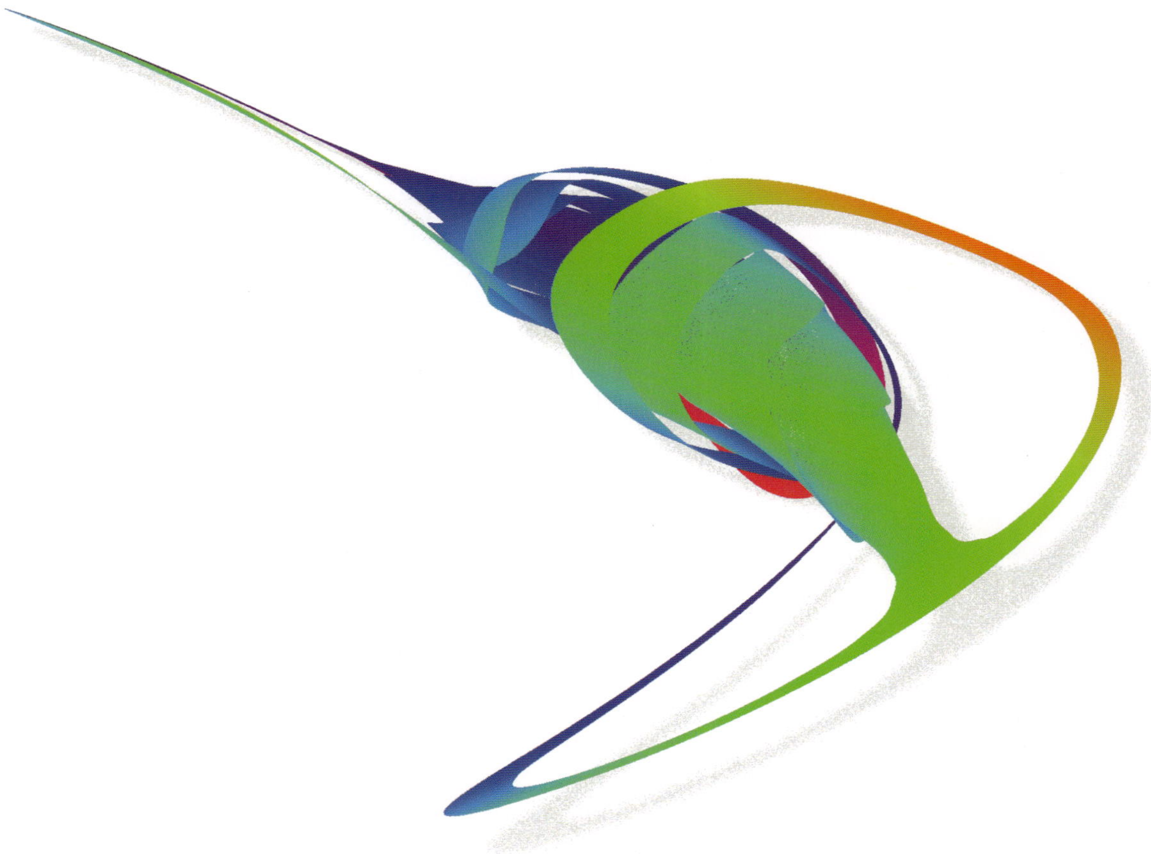

Hysteresis

A dynamical system exhibits hysteresis if increasing a parameter produces a sequence of behaviors that do not reoccur at the same values of the parameter when that parameter is decreased, and vice versa. Some people think the climate may exhibit hysteresis. If the amount of carbon dioxide in the atmosphere (the parameter) increases, the Earth warms by the greenhouse effect. The polar icecaps then melt, causing the Earth to absorb even more energy from the Sun because vegetation absorbs more sunlight than snow and ice. To get the icecaps to reform then requires lowering the carbon dioxide level to a lower value than was required to melt them. Hysteresis is a form of memory, and the hysteresis that occurs when iron is magnetized is the basis for storing data on magnetic tapes and disks. Hysteresis requires a nonlinearity in the governing equations, and the nonlinearity is often in the form of a threshold where the behavior switches abruptly from one form to another as a parameter is changed (a bifurcation).

HYSTERESIS

Whatever we expected would happen next,
it wasn't this — the error in our thinking
propagating exponentially as time unfolds —
just a little rise in temperature and the glacial ice
calves icebergs, ice shelfs, slips seaward
riding the heat of its melting. Cool fresh water
bearing the Arctic precipitate of all
our breeze-blown hydrocarbons, PCBs —
undrinkable, as we thirst in tropical heat,
the very ground sinking beneath us —
or is it ice, glaciers reversing their retreat?
And the way back to today, if there is a way,
far longer than we think. Which once we knew:
an ounce of prevention worth a pound of cure.

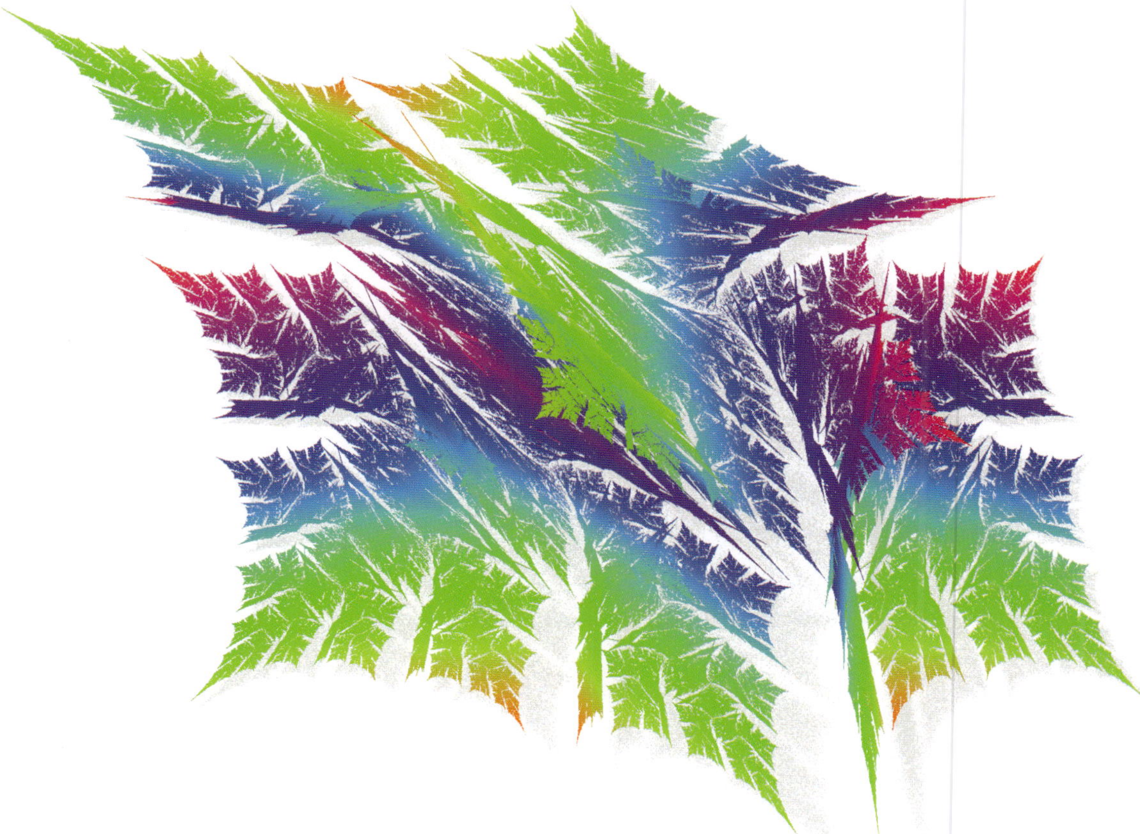

TIME, NOW

for Becky

Time, now, for the mothers
to bring flowers to each other,
to look deep into each other
to hear of daughters beginning lives

I see how you once might have become
a doctor, moving through wards
in a poor country, listening to hearts,
taking histories, prescribing cure

how you listen deeply in your poems,
how you speak the anger and hurt
of those who don't know what is wrong
how fiercely you make the medicine

that could make them whole and well
how your daughter, now, speaks fiercely
of her plan to be a doctor, how you
have taught her doctor father to hear

his daughter, so that, at Christmas,
he sends her a stethoscope
that will reflect in polished steel
the blue of the flowers she sets

by the window. How the girl you were
might have become a doctor, if
your doctor father could have heard
you. Time, now, to bring flowers

to each other, to look into the other's
life — I see you now, teaching us
to listen to the rhythms of the heart,
to the irises' deep blue.

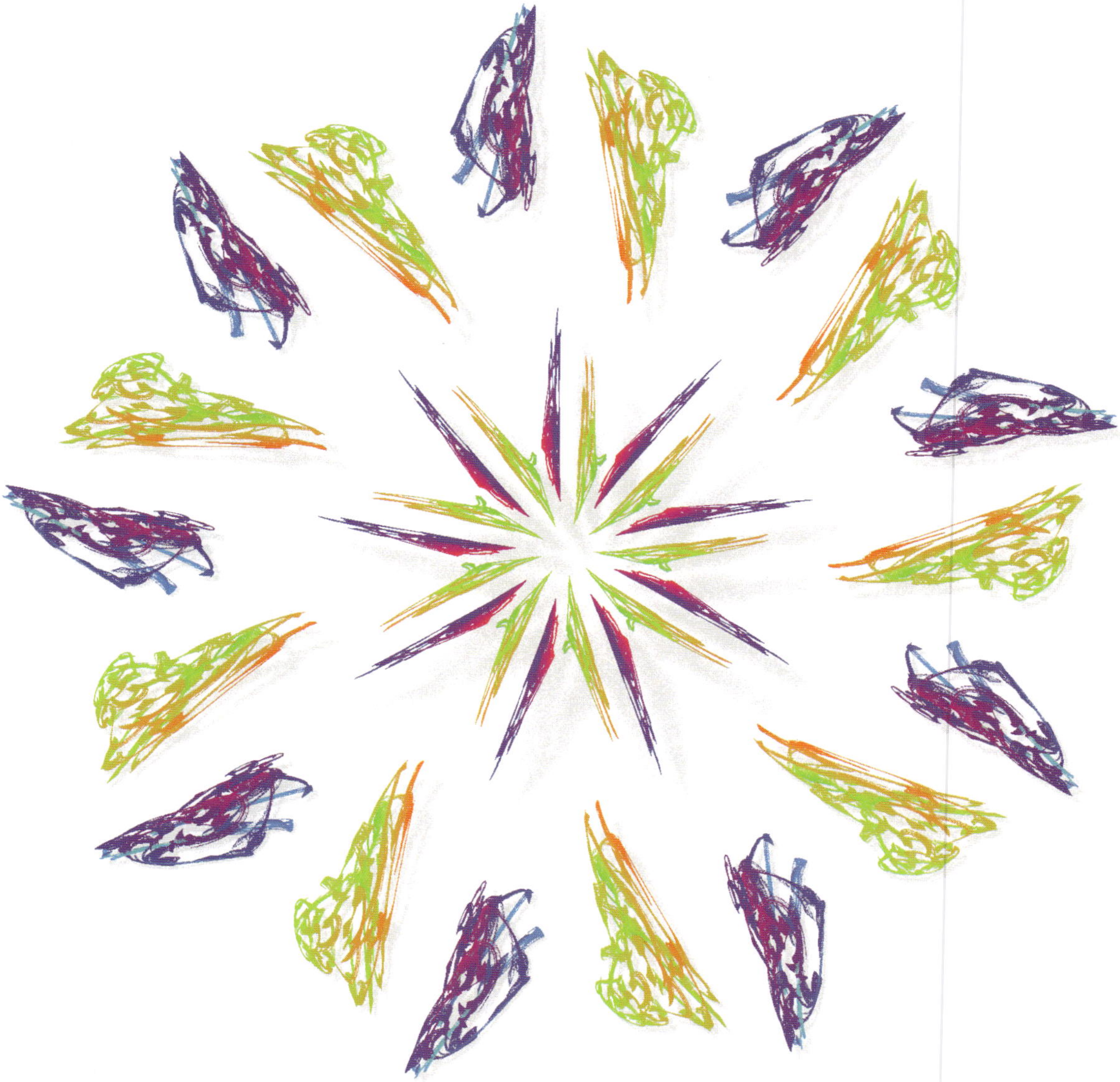

Iteration

Iteration is a kind of "mathematical feedback" in which the current values of the variables are used to calculate the next values, which are then fed back into the equations to get the succeeding values, and so forth. The result is a sequence of values (a time series) representing the temporal behavior of the corresponding dynamical system. The values lie somewhere on the attractor after a sufficient number of iterations for a deterministic system. The prediction becomes less accurate with each iteration for a chaotic system, but the solution will lie somewhere on the corresponding strange attractor. All of the images in this book are produced by millions of iterations of a dynamical system described by a set of equations.

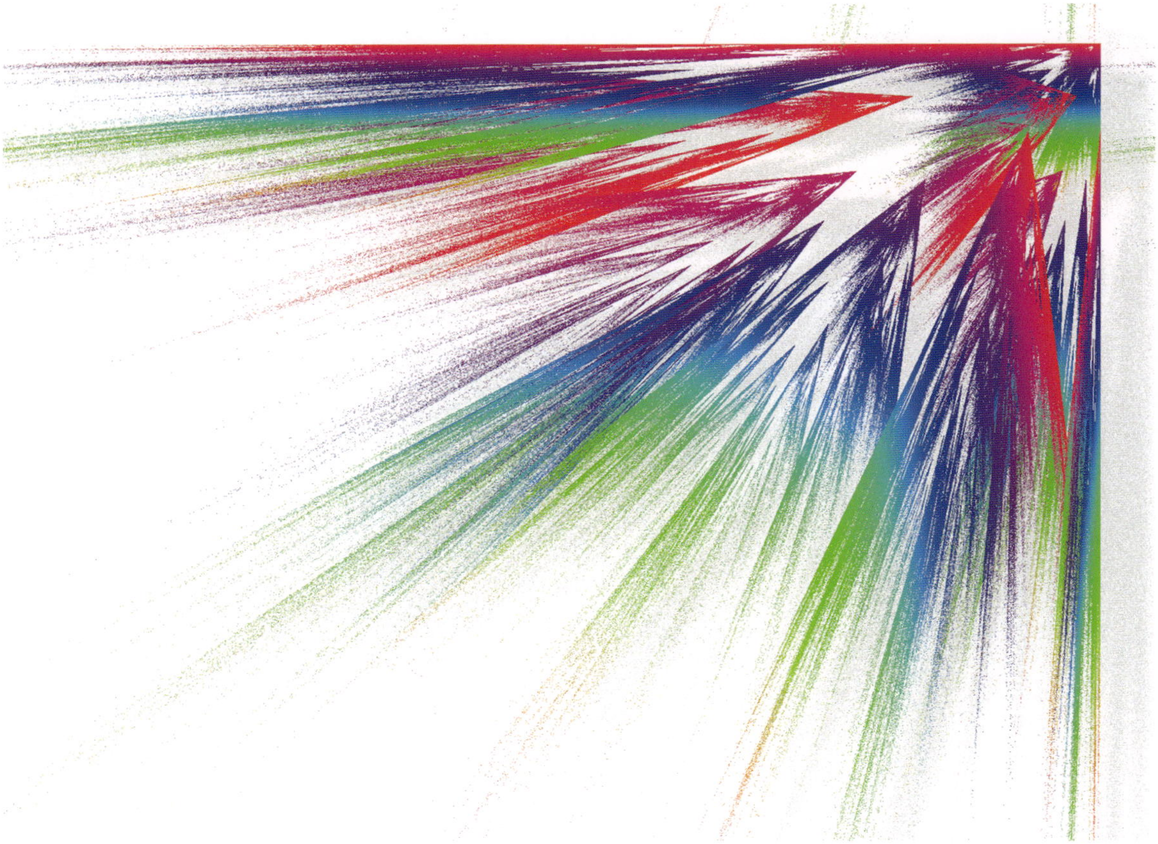

BEFORE THE DOG SAW A SQUIRREL

I wanted to say what the light was like,
How that evening it caught in the tops of the maples
And set them glowing green and orange, how the blue
Sky showed through the top bare branches,
And in the west there was a pearly white radiance
That was a cloud bank, feathered with silver.
And where I stood in the street, in shadow,
The sun just touched the burning bush, *euonymous alatus*,
A transparent rose full of the voices of sparrows.

It seemed I could stand there forever,
As though my flesh too were transparent,
The leaves of my body shaking and shining,
Green going to orange; the sky falling
Blue and white through my eyes; and the burning
Bush speaking with the tongues of sparrows
As their voices rose up in me, praising.

CANOEING, EARLY APRIL

A squawk and the Canada goose
We didn't see flaps from her nest —
Bent cattail stems, white down,
The large pink-mottled eggs. We circle out
To offer room, scolded every fifty feet
By the red-winged blackbird males.
Each drop in wind brings the insect swarm —
Midges or gnats or tiny flies in zigzag clouds.
Spring is unmistakable at last
In mallard pairs, the reddening of dogwood stems,
The loud burbling baying yelping calls of cranes.
Beneath the rumpled surface that was ice
Last week, deep down in the liquid surge
And bottom-muck, the first waterlily pads
Unfurled fresh green, began their climb.
Maybe peace, too, begins that way,
Deep down in broken things.

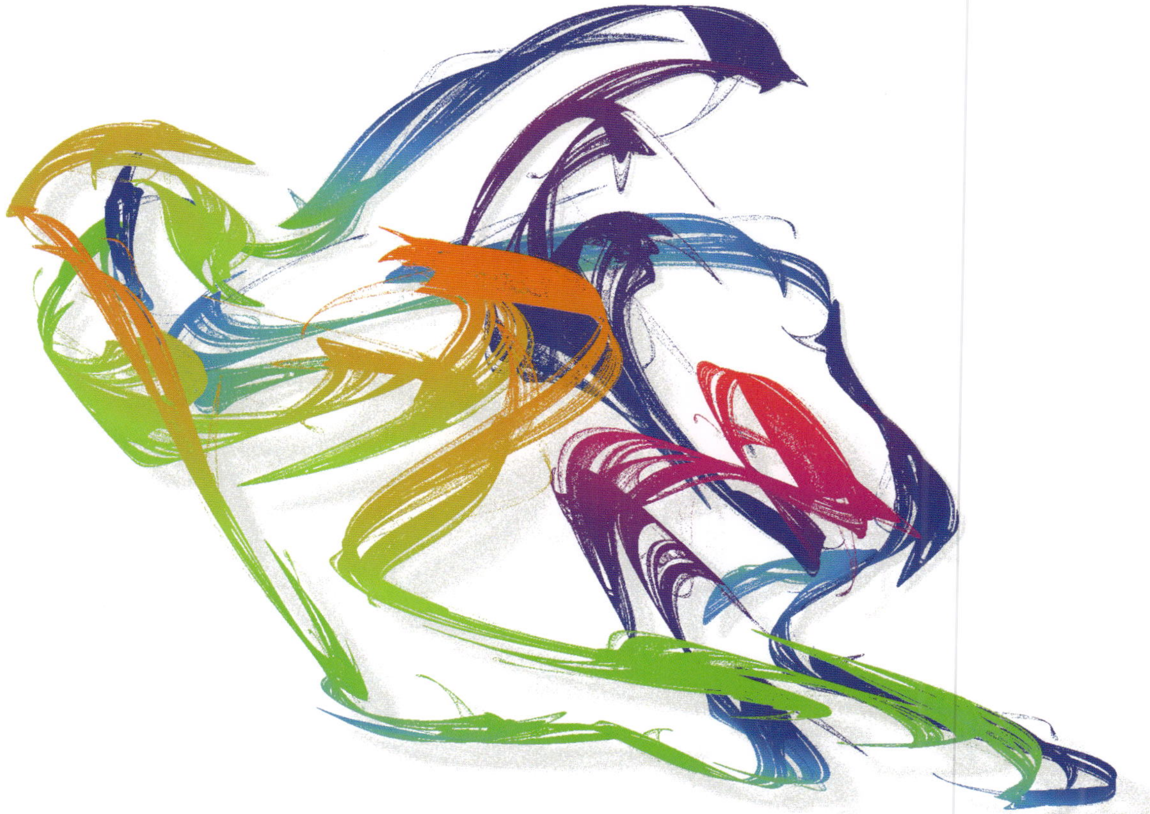

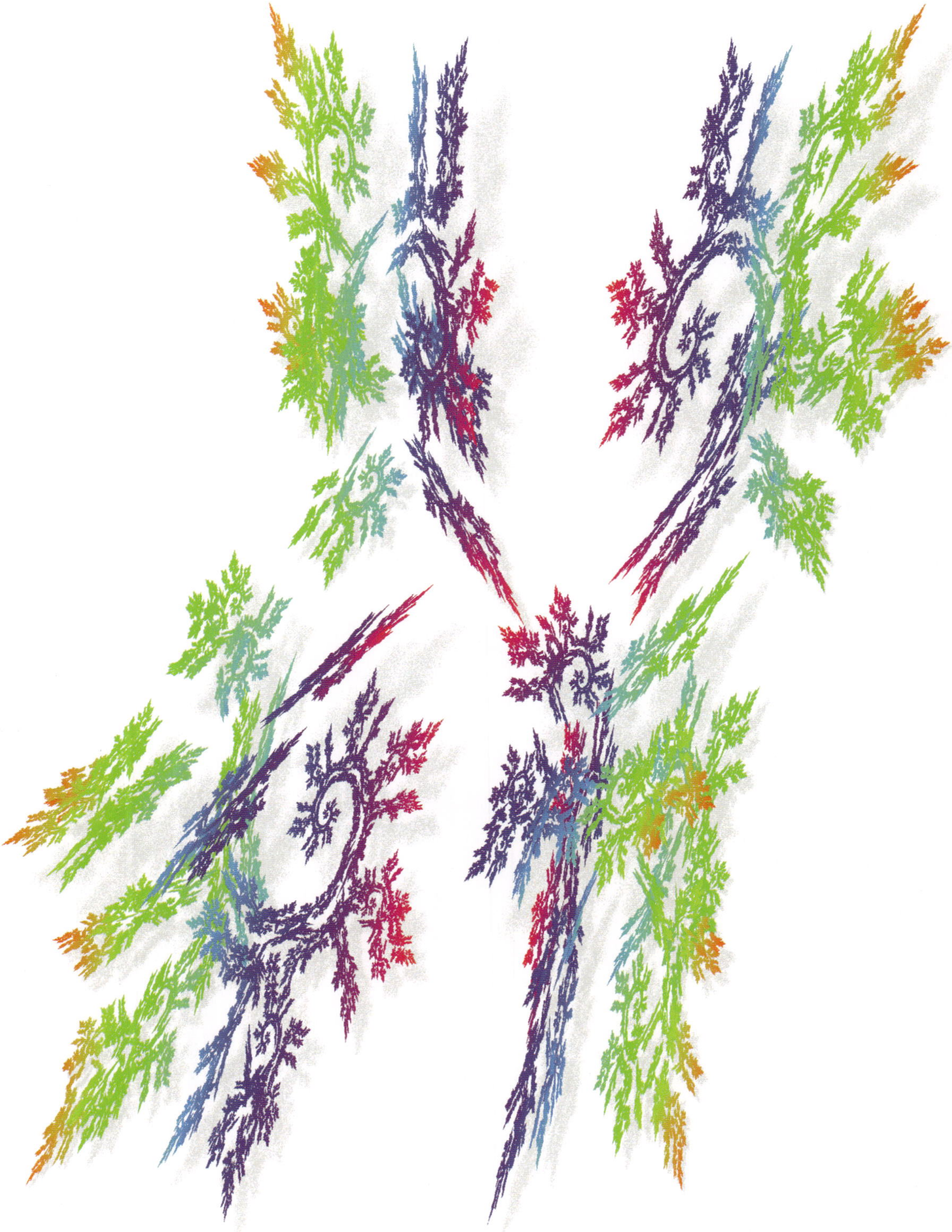

BRINGING ORDER OUT OF CHAOS

for Monica Schokkenbroek

True, we never step in the same river twice,
the flux of the universe evolves, evolves,
but any paddler or graphic artist knows that sight
can order the riffles of rapids, the shadowy waves
of underlying rock, the evidence along the bank
of flash flood, spring river height, otter slides,
the signs of beavers at work — Monica teaches me
how place and color and shape can trace
the order that recurs, recurs, in the flow of a book —
the artist pulling her own strange structure
out of the chaos of another mind;
the photographer who sees in landscape
the map of the weather of the earth.

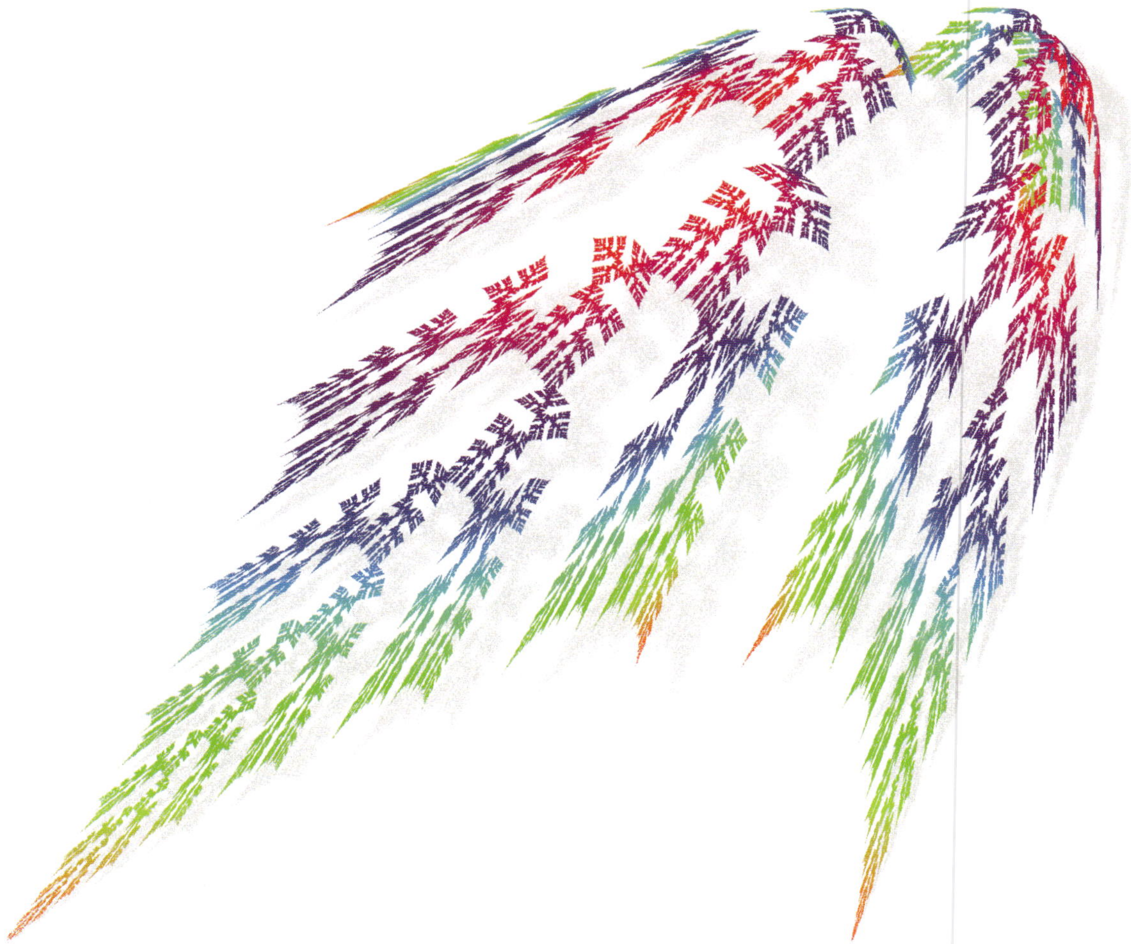

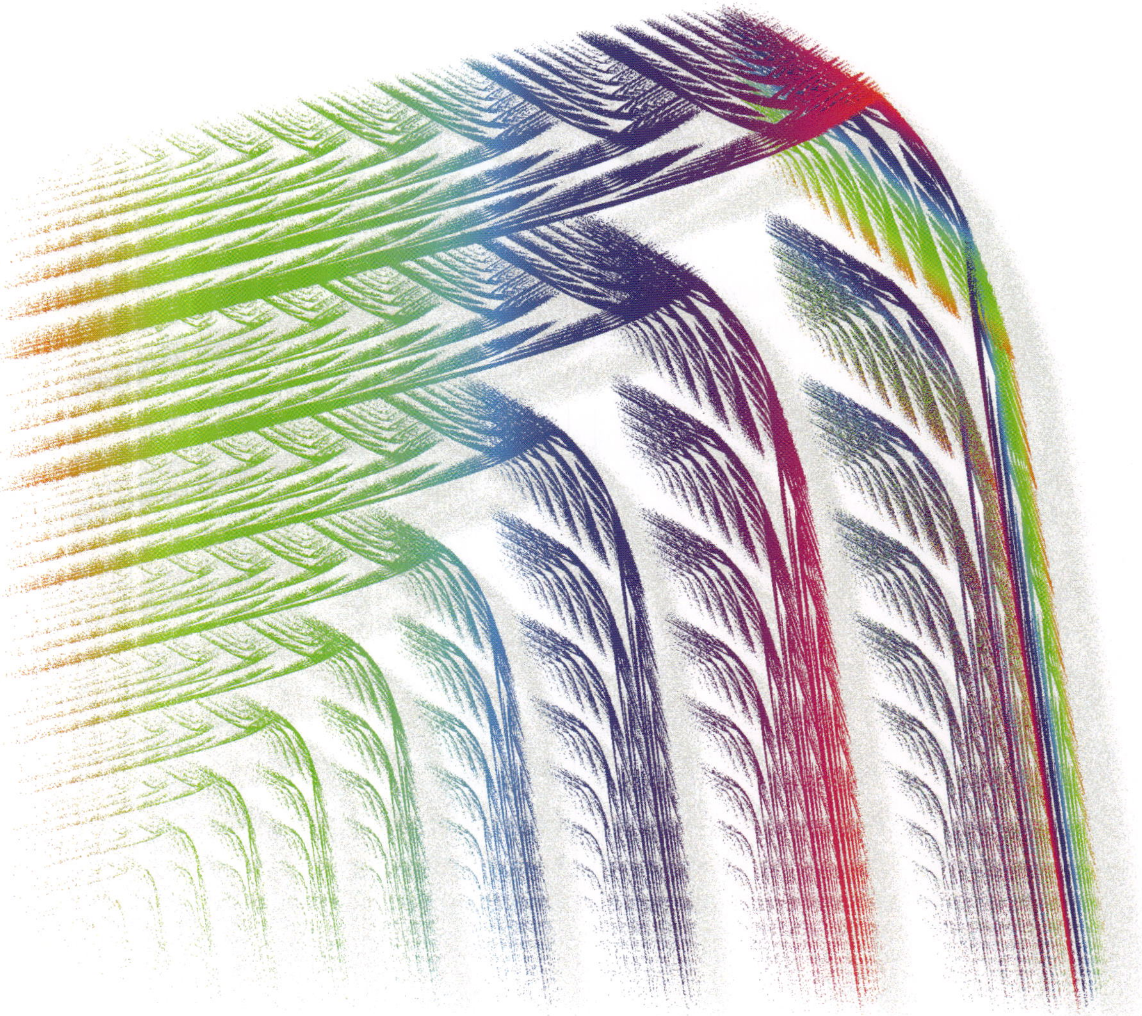

ONE LINE OF CLOUDS ON THE HORIZON

And the sun sinks through to premature departure, slips out again, a slim band of orange, square-rigged as though some mythic voyageur passed through our world, setting a path aflame to shore; and all the gulls rise up in cries, flying down that path to open water that darkens in the dying day —

And the gulls, gathered together, sleep all night in the lapping waves, wake from time to time to paddle closer to the mewling cries and pipings of each other, a floating colony under the stars. From time to time I wake, and look at the stars that stipple the sky, the lake — its waves a drowsing drift, an overfull-cup that spills no drop, so quiet the wind has grown.

Oh stars that are rushing away from us, old light of new beginnings just now reaching us, teach us how irreplaceable we are —

THE RIVERBANK

Drone of boat motoring upstream,
bright falling song of the chickadee,
bumblebees and bees humming
in the dame's rocket nectar,
young sparrows noisy
in the martin's box; geese, flickers,
crows, jays, and the wren
loud and urgent additions to wind —
only the river silent, silent and broad
and rising — the river birch
and sandbar willows half
submerged, the bank edge
altering by the hour — our Edens,
idylls, riverbanks of memory
fixed only by our brief attention
to the river's endless shifting,
falling and rising like birdsong,
or the orange butterfly visiting
dandelions, one sweet taste of life
after another, the stretches of green
in between. Oh, what shall we do
with our lives that last
a blink of the river's,
a hundred butterflies'?

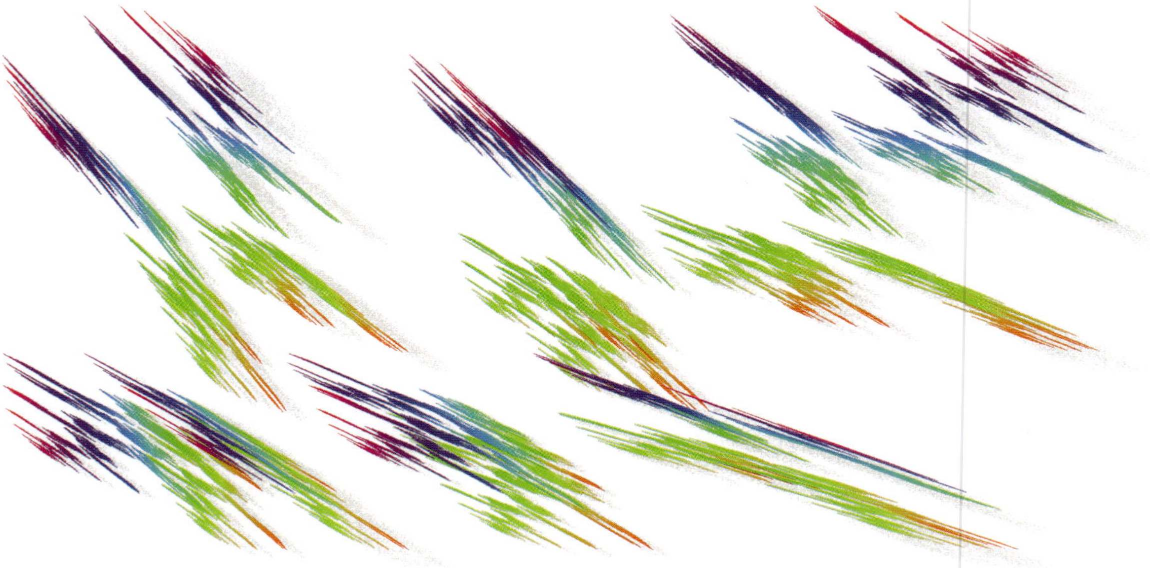

THE AVALANCHE LILIES OF PARADISE

How many spill across high slopes
stripped by glacial ice and avalanche,
bursting forth these few summer weeks
through the shrinking patches of snow —
slick green chutes of leaf, crowded,
low, tangled in subalpine meadow:
and the white-winged coifs of the lilies
themselves, a nodding choir
chiming silently the bells of Paradise
below Mt. Rainier's lenticular stack of clouds;
when these slopes are shaken to the liquid mud
of a speeding lahar, or blasted loose
in volcanic fire, the lilies will follow,
some few, to root again the ashy soil,
multiply their hold on the bare rim of the new,
needing only cold, and snow, and years —
oh hear them *leaf* and *leaf* and *flower*.

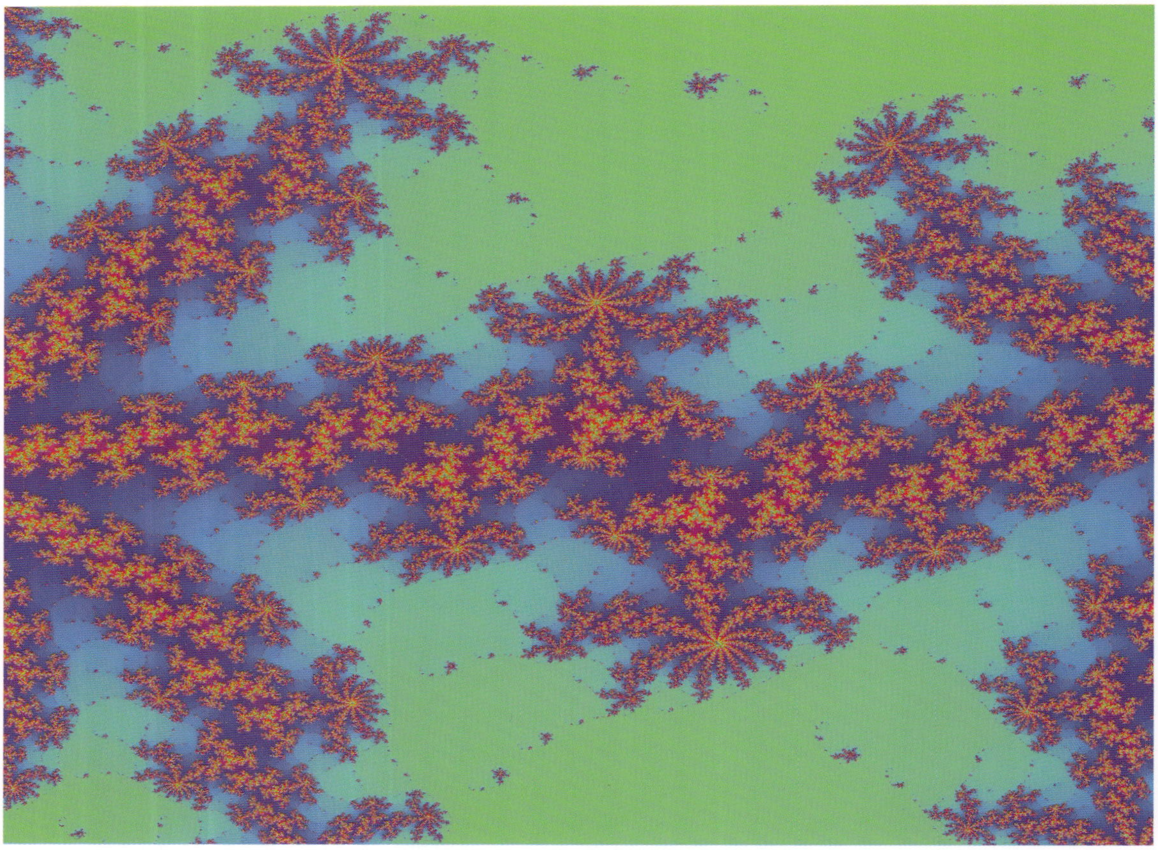

The Butterfly Effect (Sensitive Dependence)

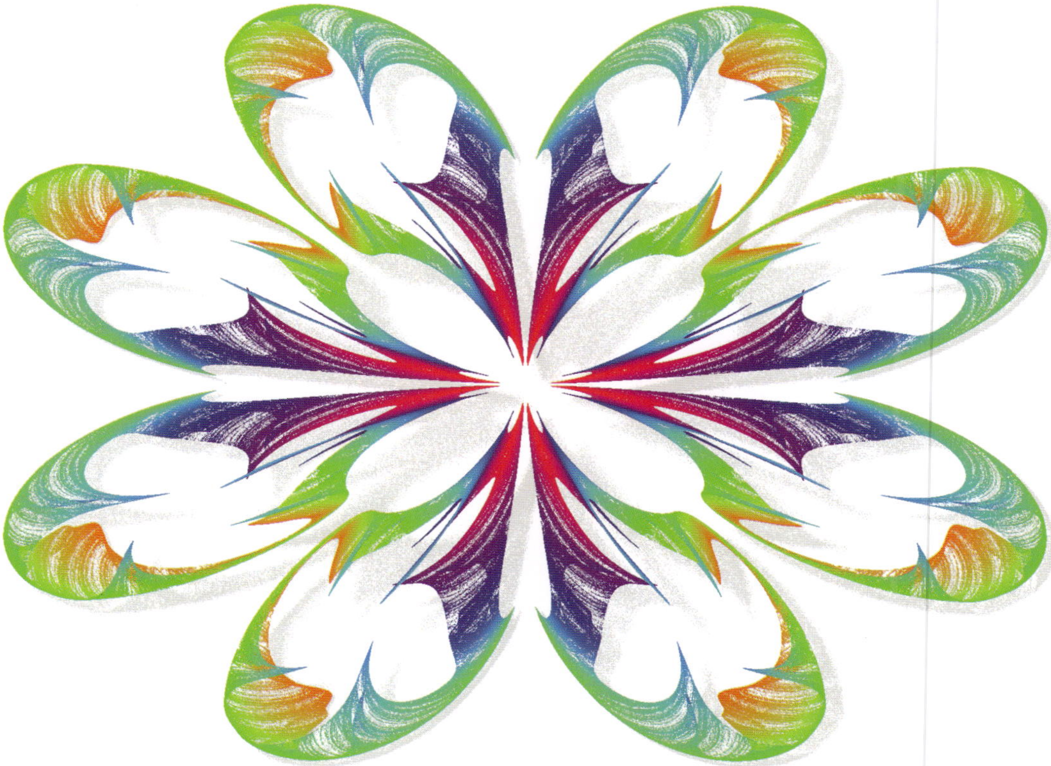

Butterfly effect

The sensitive dependence on initial conditions that is the defining characteristic of chaos is often illustrated by an example from the weather, which is determined by the motion of the atmosphere. Tomorrow's weather depends on today's conditions (the initial conditions), and the day after depends on tomorrow, but small changes in the current conditions make a larger change tomorrow and an even larger change the following day, and so forth. In a chaotic system, the error grows exponentially so that a tiny change in the current condition will eventually grow to a large size. Consequently, a butterfly flapping its wings in Brazil can set off tornados in Texas a few weeks later. Wave your hands in the air, and you can be sure the weather will be completely different several weeks from now than it would have been if you had not done so.

THE SLOUGH OF TIME

Lazily the water moves, and with it
The swirl of algae bloom,
A kind of paisley, dotted

With floating leaves, duck down,
Duckweed, snarled fishing line,
An empty bottle.

Over the slough hangs the hum of summer,
Crickets loud in the browning grass,
Blackbird call, crow, the mower.

Lazily the water moves and reverses
Past the dragonfly at her reedy post,
Her many eyes taking in

The watery universe spread below —
No unrepeatable river but the scarf
Of chaos, traced by the passage of wind

And nymph and mallard — threads raveling out
Into this particular world that could,
If she fans her wings, become another.

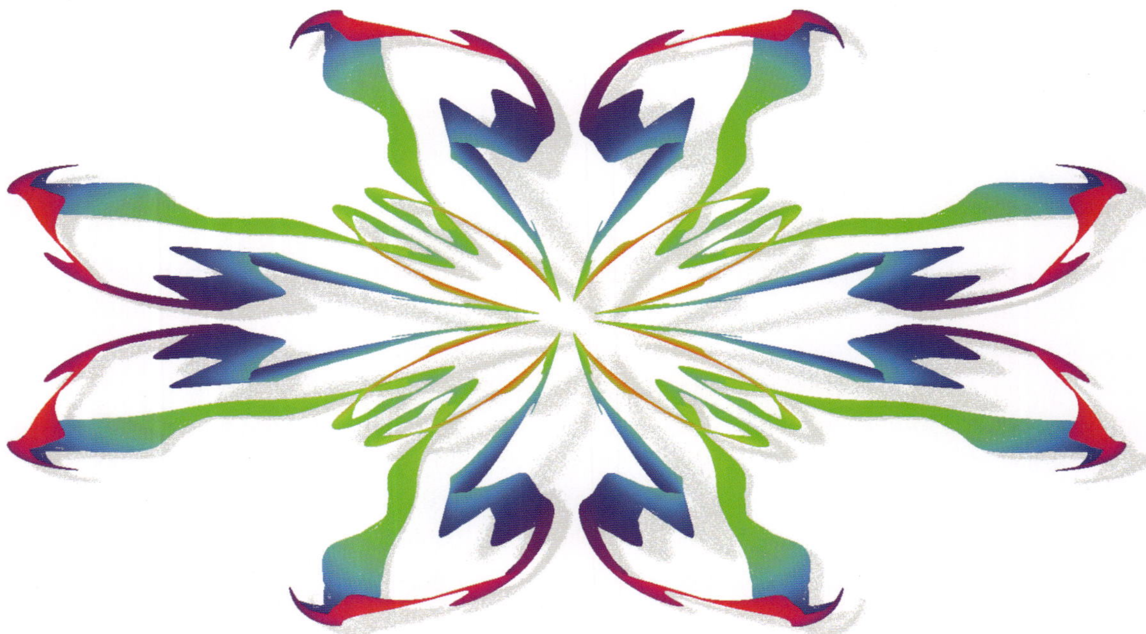

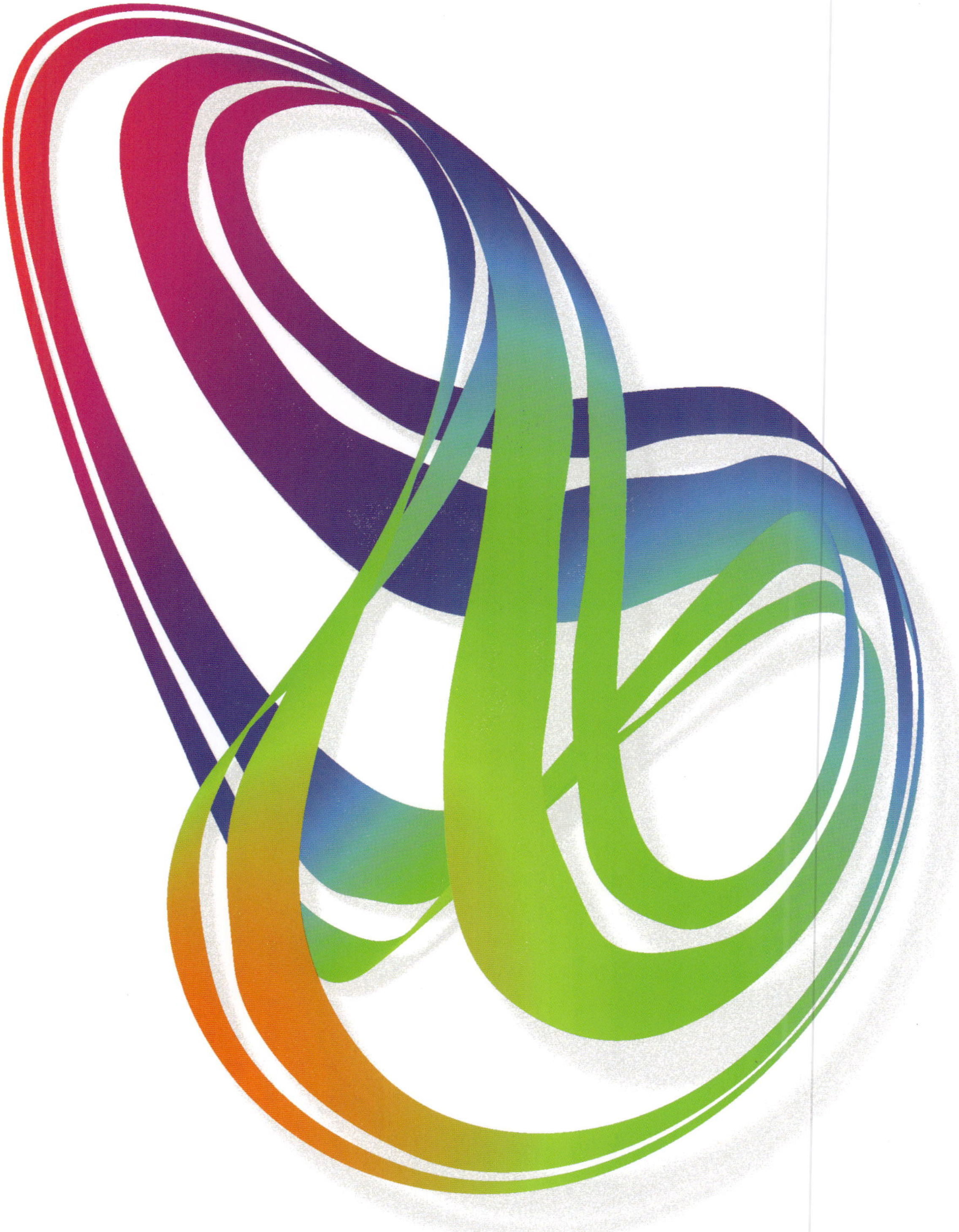

BACKYARD

I let the fence rot, the grass go;
Butternut, creeper, dame's rocket,
Cherry tree, blackberry spring up.

Yearly the flicker returns to drum
The tin chimney, the doves mourn
In the locust, the owl hunts the shrew.

In February, raccoons
Lead their young
Past the back door.

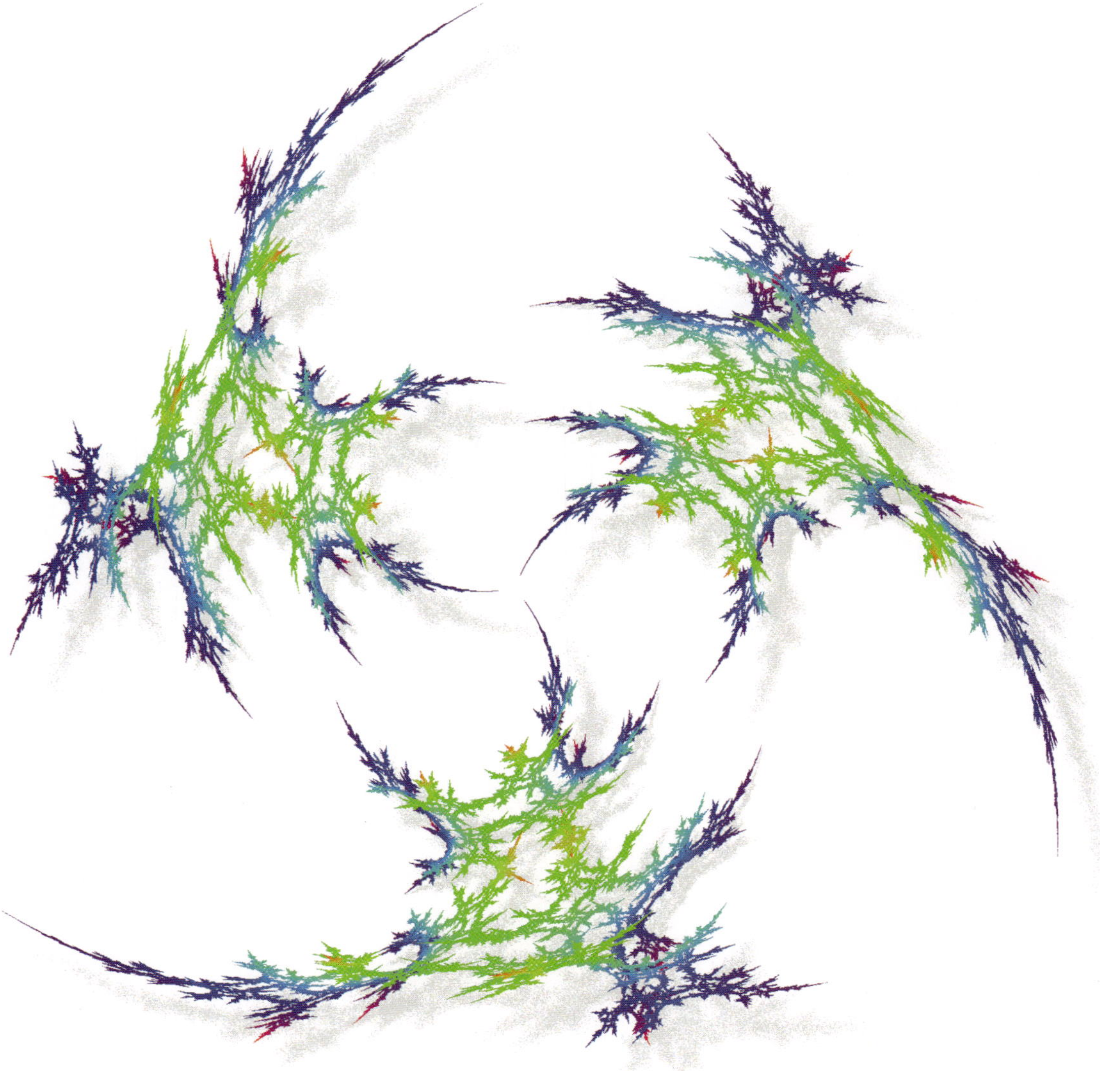

FROM THE UNION TERRACE

There's a man in a pith helmet
and beard, shirted in palm leaves and parrots,
eating a power energy bar,
looking over the lake, talking on his
cellular phone — only time for a quick call
before he darts away, his powerbook
notebook slung over his shoulder, stalking
other primate game than the t-shirted,
ragged-jeaned beer drinkers and brat eaters,
golden in their summer tans, or the pony-tailed
English profs, undershirted sailors, the table
of history majors just working their way
out of the early twentieth century, the band
cranking up amps for the twenty-first,
the end of history. A man in a pith helmet
phoning from a midwestern lake to learn if
the rainforest monkeys he studies
are safe in the season of slash and burn,
the sounds of crackling coming over the line.

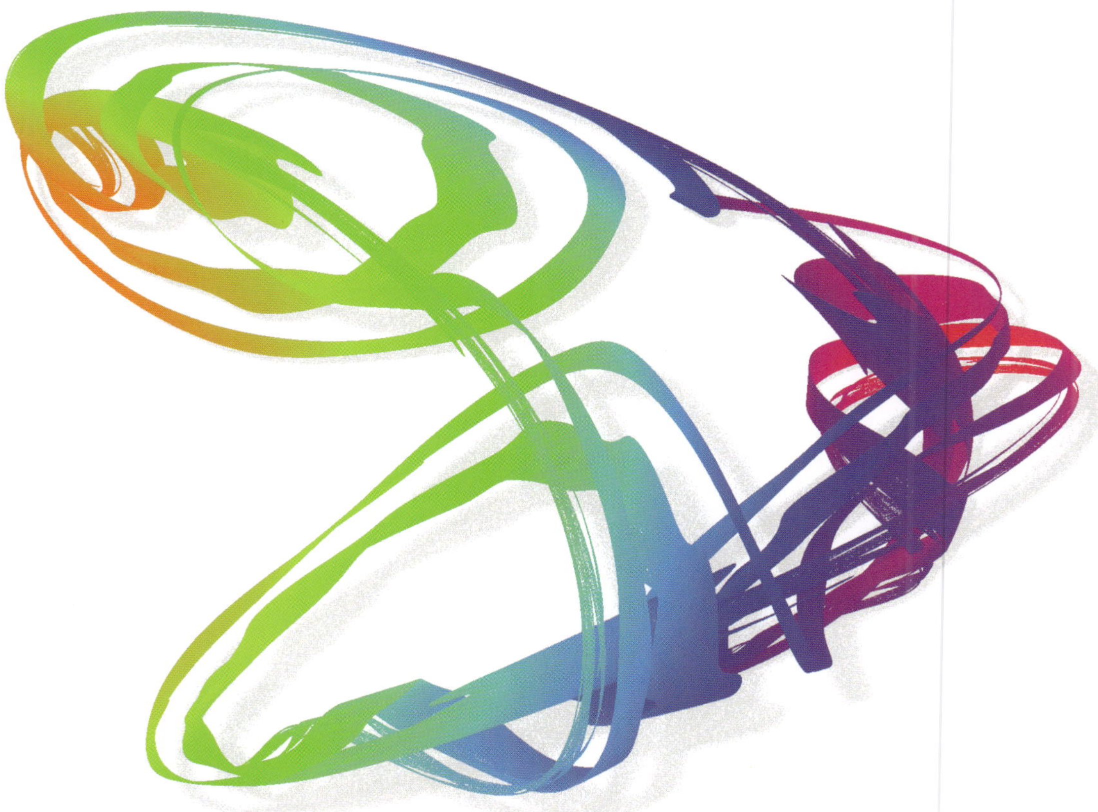

EXPLAINING THOSE DISCONNECTED CONNECTIONS

for Will

Last night, for instance, how airplanes flew
through both our dreams, yours swooping and dipping
into the lake like swallows, mine
a dark cavernous auditorium of silence
that bumped along the jungle grass,
transportation I almost missed;
or those startling conjunctions in life,
how at manuscript group one Wednesday
Art and Jean could both bring
stories to the group
about castrating sheep —
those resonances of the world net
that Jung named 'synchronicities',
that physicists protest, suggest
their own quantum theory of action-at-a distance.
Biologists say its something we ate
and psychologists attribute it all to the stories
we make up after the fact,
though historians might cite
something in the ether called Zeitgeist —
but those swallows, I saw them too,
canoeing with you,
on the banks of the Green River, Utah.
They scooped up water and mud,
returned to plaster those nests like breasts
on the red-rock cliffs, and that auditorium —
you've been there too, sitting in the back,
listening to me speak my poems into the dark,
and the airplanes — what are they but wings
that will carry us into our lives together?

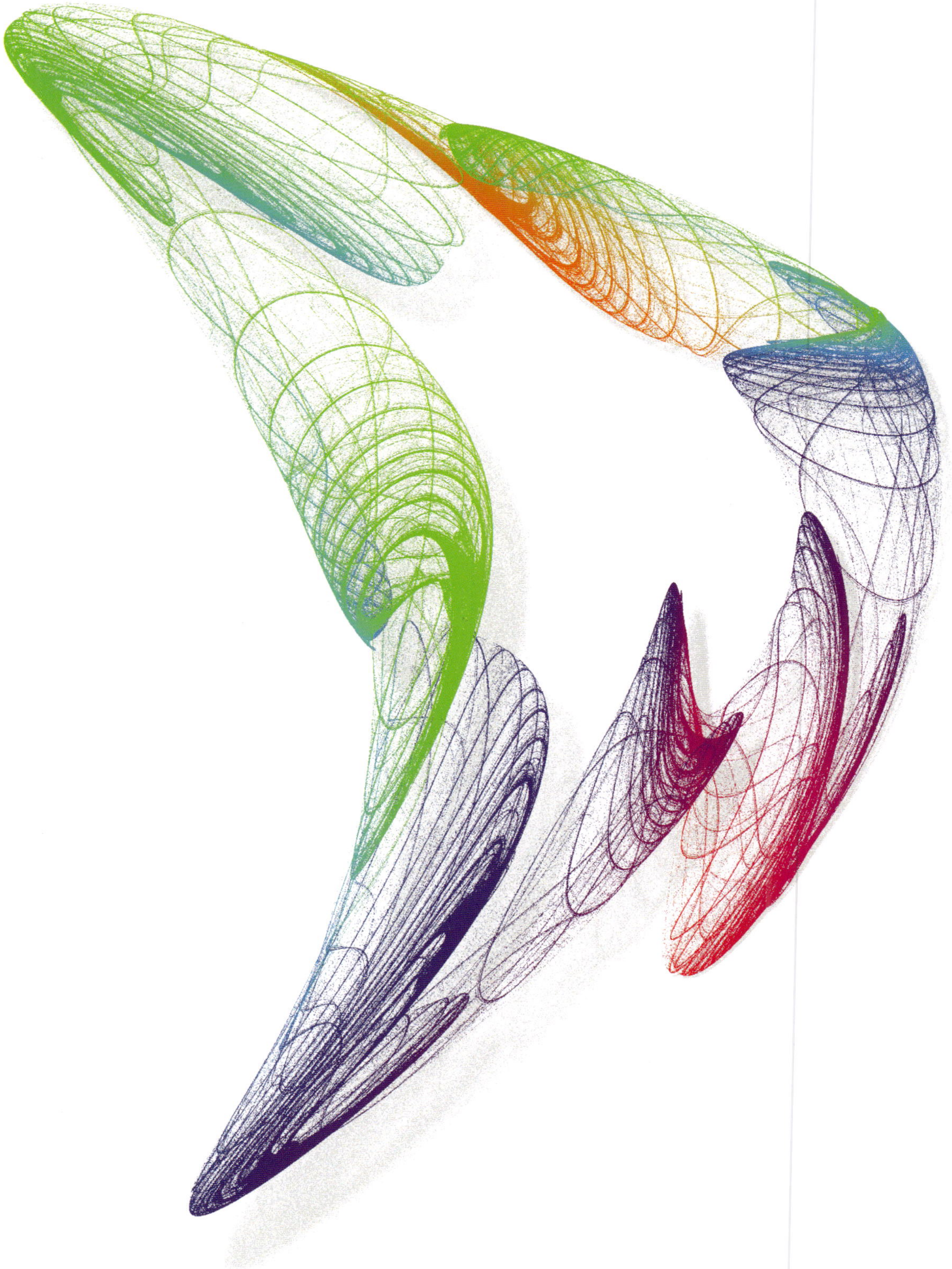

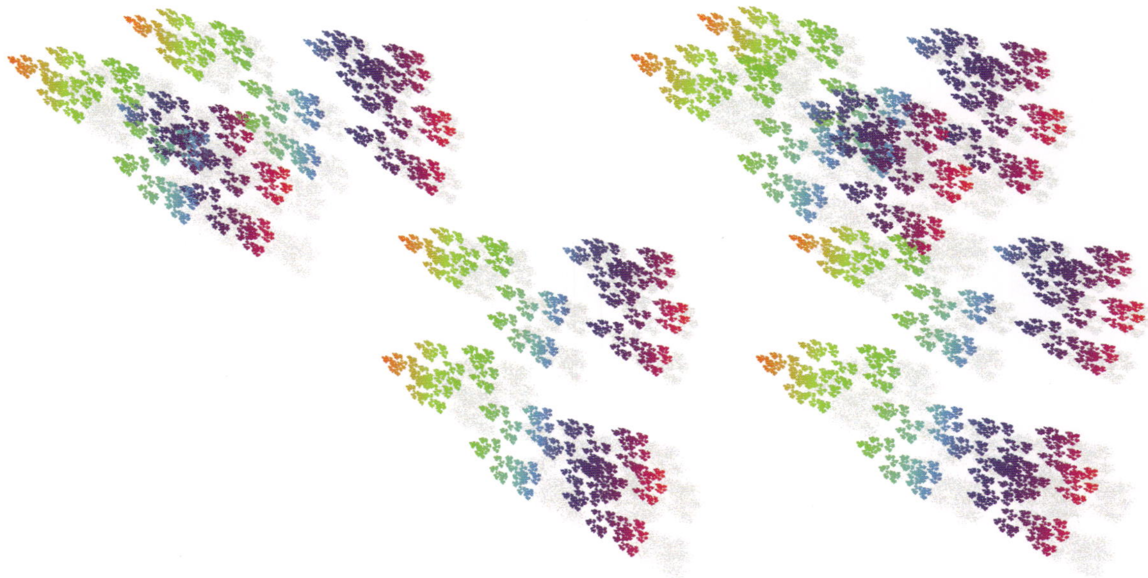

ARRIVING

In a field winnowed of color, brown grasses and stubble
of milkweed stems spilling their seed and silk, we look up
into a sky of gray clouds, tendrils of mist moving over;
look up into the gray billowing shape-shifting masses
where a white diffuse light moves and plays as the bugling sounds —
burbles like water running through the rocks, but louder,
arriving from miles away, from throats whose resonant vocal tracts
coil five feet of sound into their breastbones, next to their hearts,
travelers gathering here, wherever the continents have drifted,
these sixty million years, meeting before the fall migration.

They are calling, through the mist, family to family, parents
to streaky gray young, they are gathering, twenty thousand or more,
into our sight, great necks outstretched, adults cockaded in red,
long legs clumsily angling down, as one flight after another comes in,
hovers over the dun field, each sandhill crane back-flapping its wings,
dropping down, down onto its long dark legs, disappearing again
into the multitudes gathered here, tall presences, motionless
sticks and grass — only here and there the opening wings of the young,
greeting each other in the courtly strut of the mating dance. The calls
still bugling overhead. Ground beings, we stand, mute witnesses.

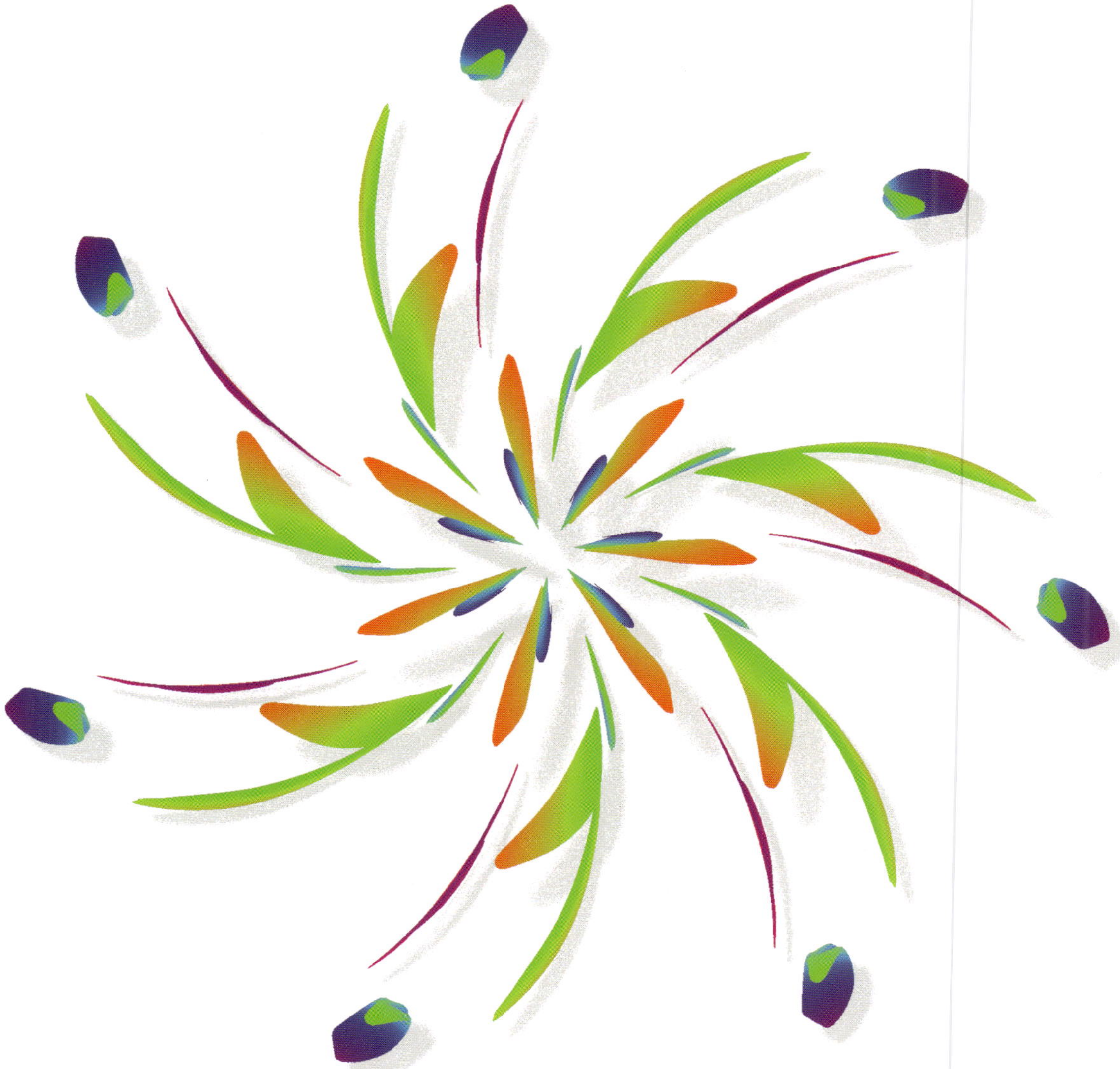

Chapter 7.

Truth and Beauty

If truth is beauty, how come no one has their hair done in the library?
—Lily Tomlin

Fractals

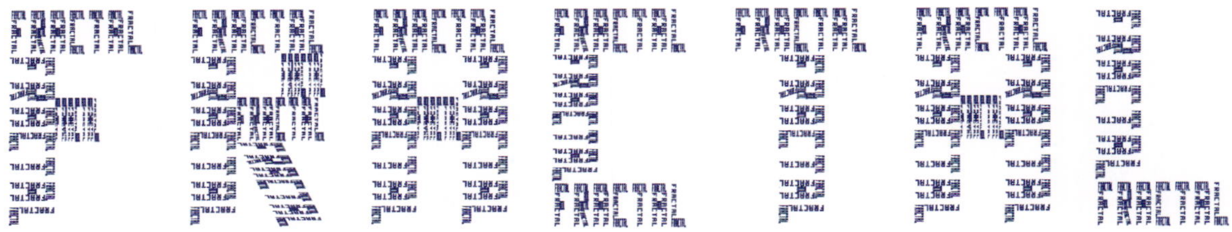

Fractal

A fractal is a geometric object that contains miniature copies of itself on all scales. We say it is "self-similar" since pieces of it are similar to the whole. Fractals can be either exactly self-similar or only statistically self-similar. They have structure on all scales, in the sense that you can zoom in on a portion of one with arbitrary magnification and still see structure. Many objects in nature such as clouds, plants, rivers, and coastlines are approximately fractal. Fractals can be characterized by a dimension, which is not usually an integer, but rather a fraction like 1.5 or 2.2, since they are neither lines, nor surfaces, nor solids. The trajectory of a chaotic dynamical system is usually a fractal, and fractals have been called the "fingerprints of chaos." The artistic images in this book are examples of mathematical fractals.

FRACTAL DIMENSIONS

The piano packs up the trees
    into keys and sounding board.
The player unfurls them as time
    and something more — echo
Of spring in the forest, the way
    that branches break into leaf;
Flood, those eddies and whirlpools
    in the turbulent river;
Voices, those spirits abroad
    in the winter storm.
Listen, the pianist whispers the name
    of his new wife to the strings
And they echo back *Rose*, *Rose*
    in an eighteenth century voice.

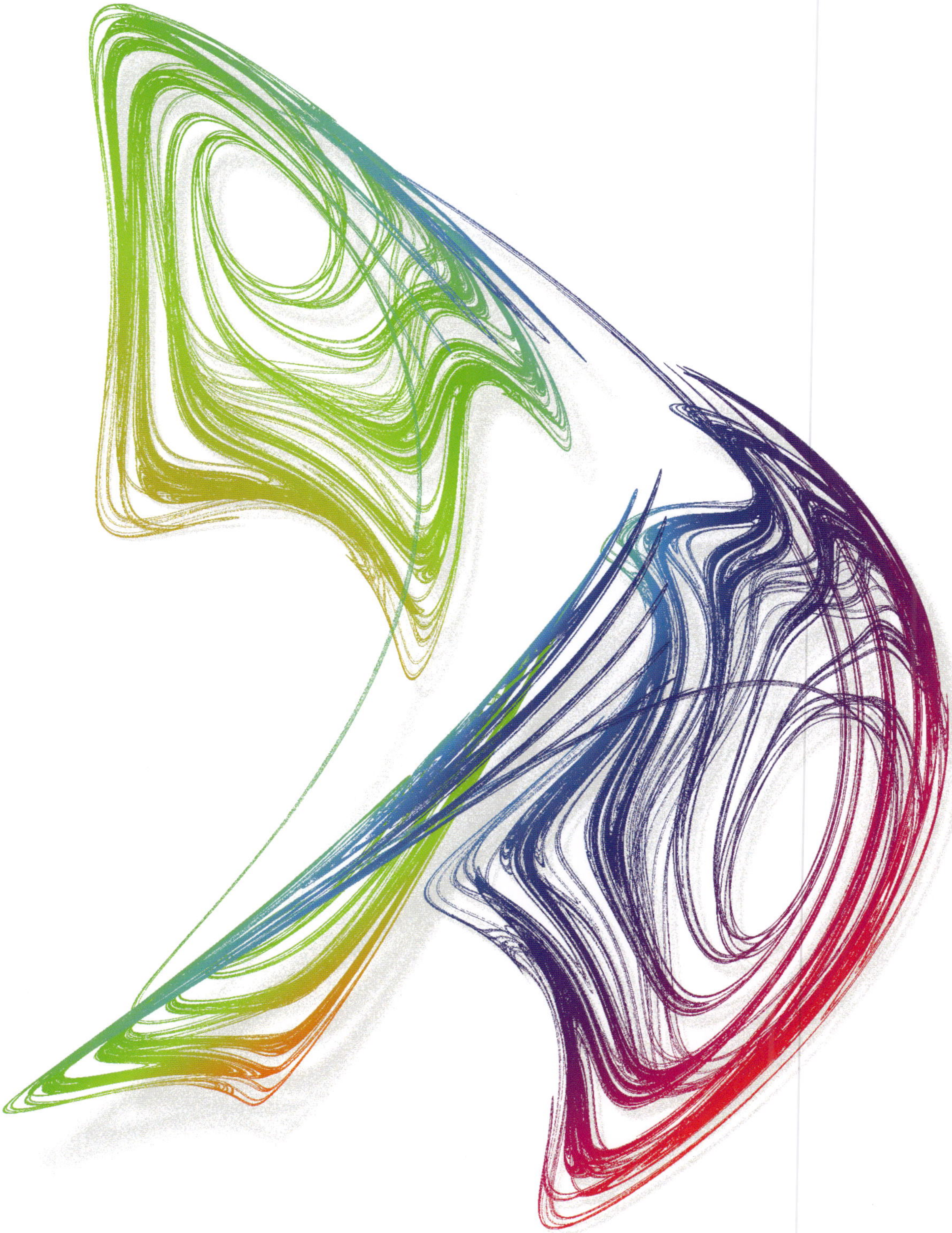

Cantor set

The Cantor set, named after Georg Cantor (1845–1918), is one of the simplest mathematical fractals. Imagine taking a string and removing the middle third, so that you have two strings, each with length one-third that of the original. Then do the same for each remaining piece so that you have four strings each with length one-sixth that of the original. Keep dividing the strings infinitely many times. The resulting object is more than a set of points, but less than a set of lines. In fact, it has a dimension of 0.631. The Cantor set is a set of measure zero because the total length of all the pieces is zero, even though there are infinitely many of them. When it was invented, it was considered a mathematical curiosity of no practical interest, but it turns out that dynamical systems that exhibit chaos often give rise to Cantor-like sets of points.

The triadic Cantor set

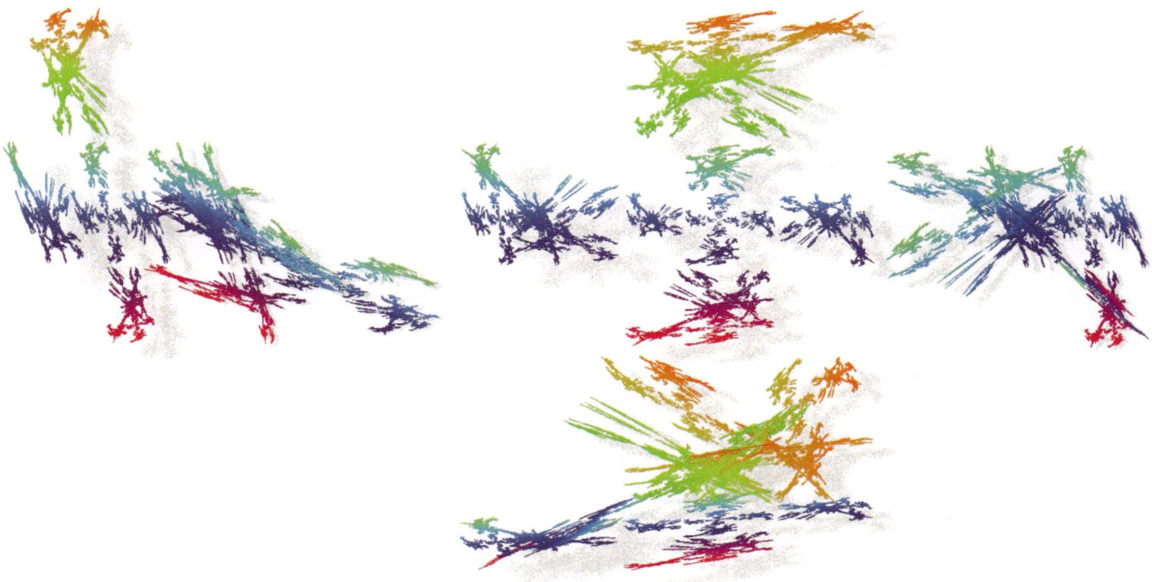

JACKSON POLLOCK PAINTINGS HAVE A FRACTAL DIMENSION OF 1.7

So says Richard Taylor, dividing photographs of yards of Pollock's canvas into smaller and smaller boxes, counting the complexity of drips and splatters and expansive gestures, a constant across four orders of magnitude — dense thicket almost filling each square from smallest to largest, a tangle like brambles or branches. No openness of unraveling clouds for Pollock, no leisurely ripple of light in moving water repeated larger and larger. The 1.7, a number to tell Pollock's work from fakes even with no signature.

Scientists say that goldfish, too, swimming their bowls, show their own fractal signatures — this one 1.5, that one 1.7, in how it hugs the wall, slows down, speeds up, darts across the space of the possible. Though they say we prefer to watch the 1.3 of clouds and waves, sand dunes' undulations, to the impenetrable 1.7 of dark woods like Dante's.

Then what was Pollock trying to do, throwing automotive paints around in arcs that broke art's canons, allied paint with dance? He held the brushes like sticks or pens and traced the air — what he saw the only image, and paint flew everywhere, dappling his shoes, the floor, the canvases he'd sell for a million. Years later, sodden with drink, he drove his car, his lover, her friend, hurtling off a curve into the dense trees of Easthampton. Under his bed for tourists to see, his shoes gaudy with drips.

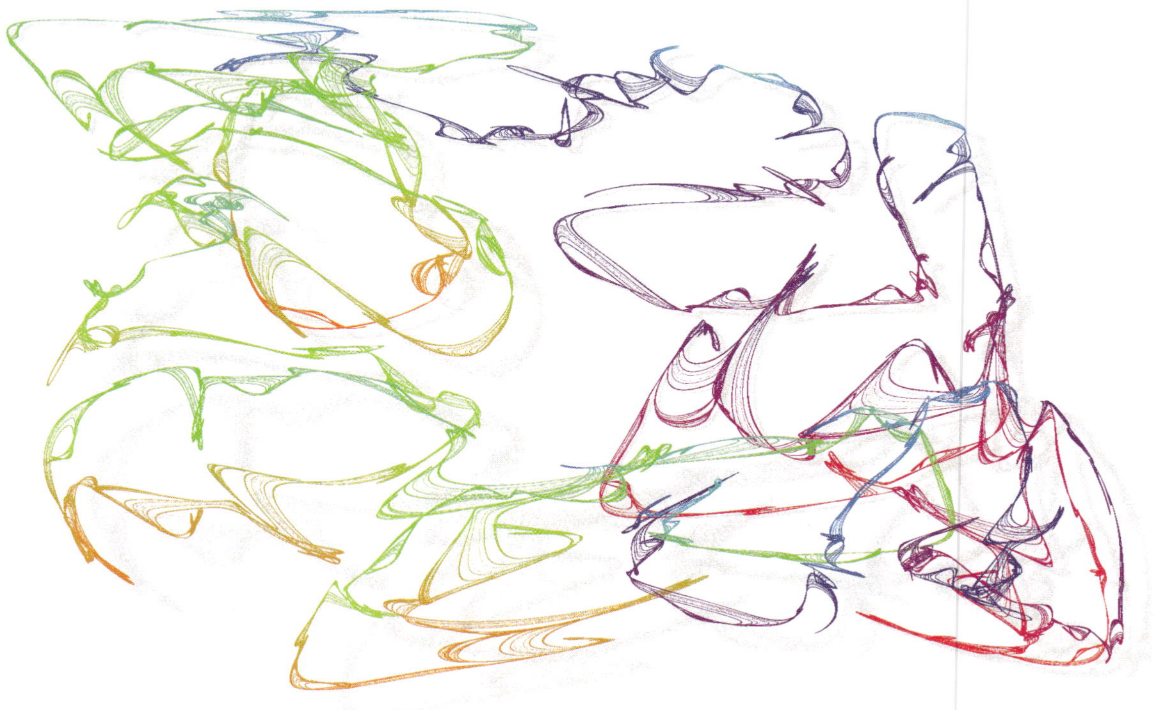

ON THE SHIFTING NARRATIVE POINT-OF-VIEW

Light makes a metaphor of lake —
How light may be refused, or focused
Like a lens, or taken in, as if the lake could choose,
As if light were the protagonist,
Lake the friend or foe.

Lake showing how
Light may travel dual, particle or wave,
Even as the literal medium bends and flows.

And water's everywhere,
Said the physicist, learning to make
Trenches one atom thick in films
At nanoscale
With the scanning tunnel microscope —

Films that first he had to bake and bake
To drive the water out,
Creating ditch and bridge on a scale where light
Like grit no longer shone but hit,
And the story was of logic circuits.

Zipf's law

The Harvard linguist, George Zipf (1902–1950), noted that if the words in English written text are rank-ordered and if the frequency of their occurrence is plotted versus the rank on a double logarithmic scale, the resulting plot is a straight line, implying scale invariance. He and others went on to observe such scale-invariance in many other quantities such as the length of words in written text, the population of cities, and the number of links to pages on the World Wide Web. He proposed a principle of least effort as the explanation, but the cause of this common behavior is still not well understood, and may point to a deep underlying and unifying principle of how complex dynamical systems self-organize in nature.

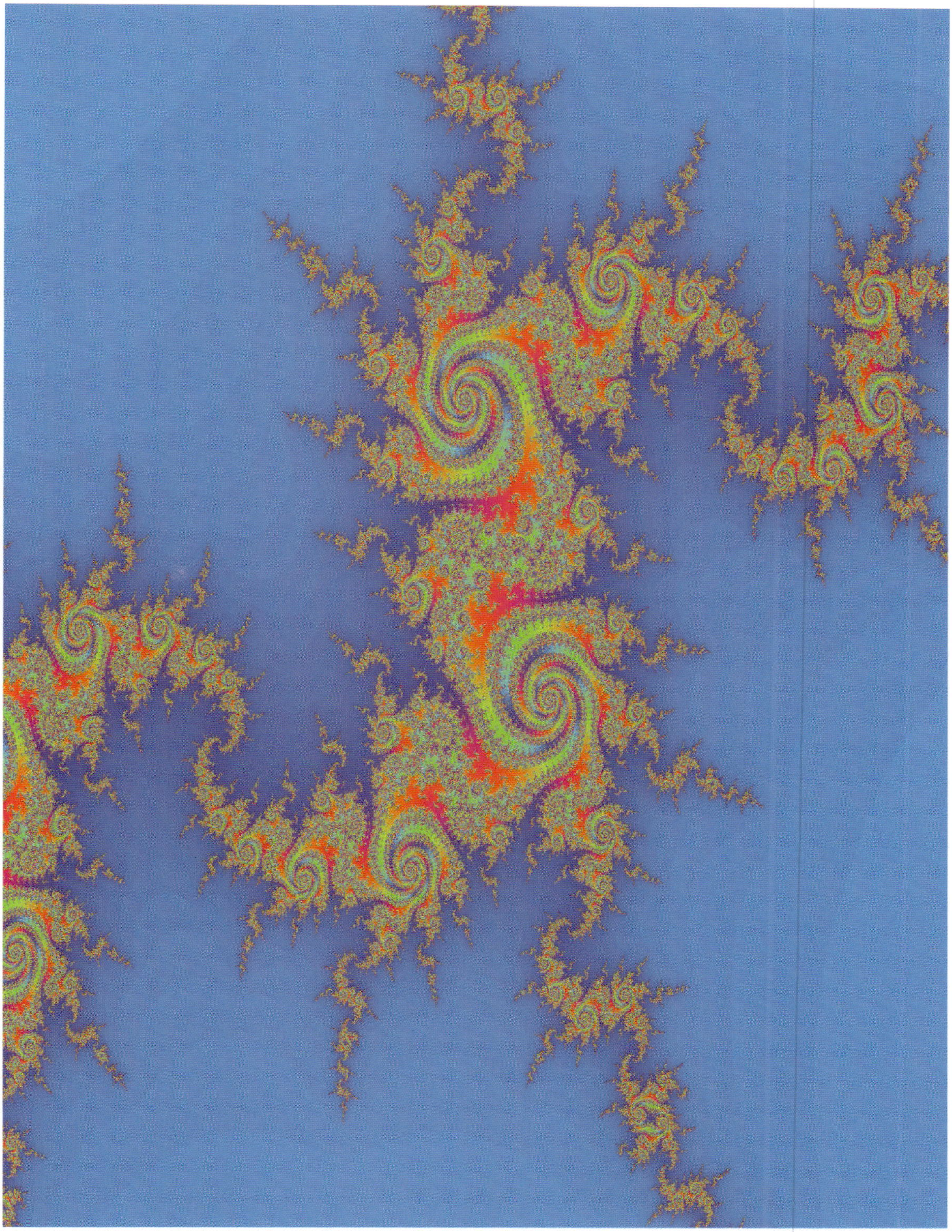

Mirror Images

LATE

Late; the fireflies
Have reached the trees,
Where they wink on
In velvet dark —
The leaves a softer black
Than sky —
And slowly off; in Java,
In the mangrove trees,
One to a leaf
All up and down river
They flashed in synchrony
Like schools of fish
Turning in unison
Or girls at school,
Monthly cycles tuned.

Here it's call and answer;
As each turns up its voltage,
Brief electricity
Deepens the night,
Intense, gold,
Glowing, gone; deep
In our bodies we answer
As though each
Were a summons,
A bell note struck —
And its resonance.

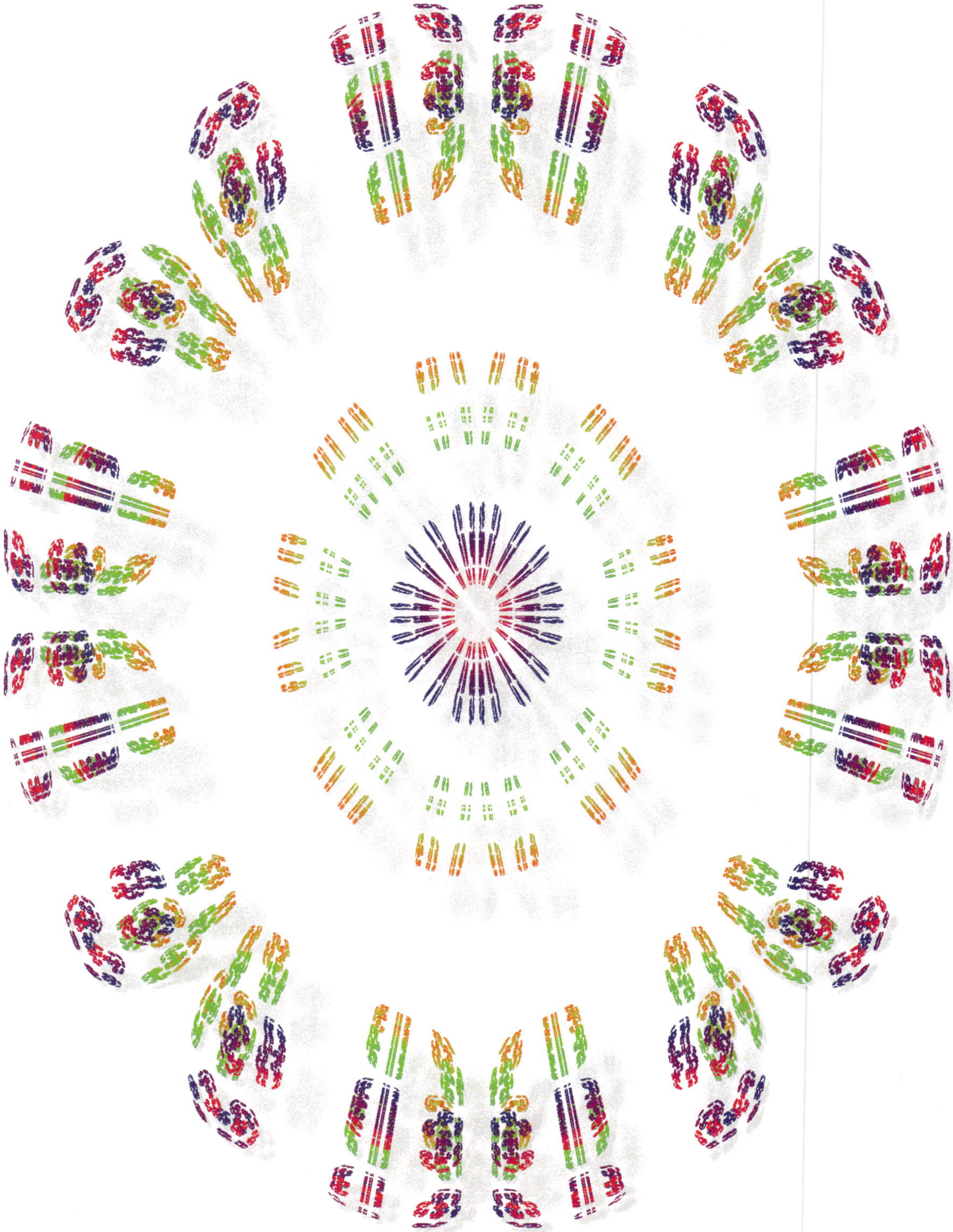

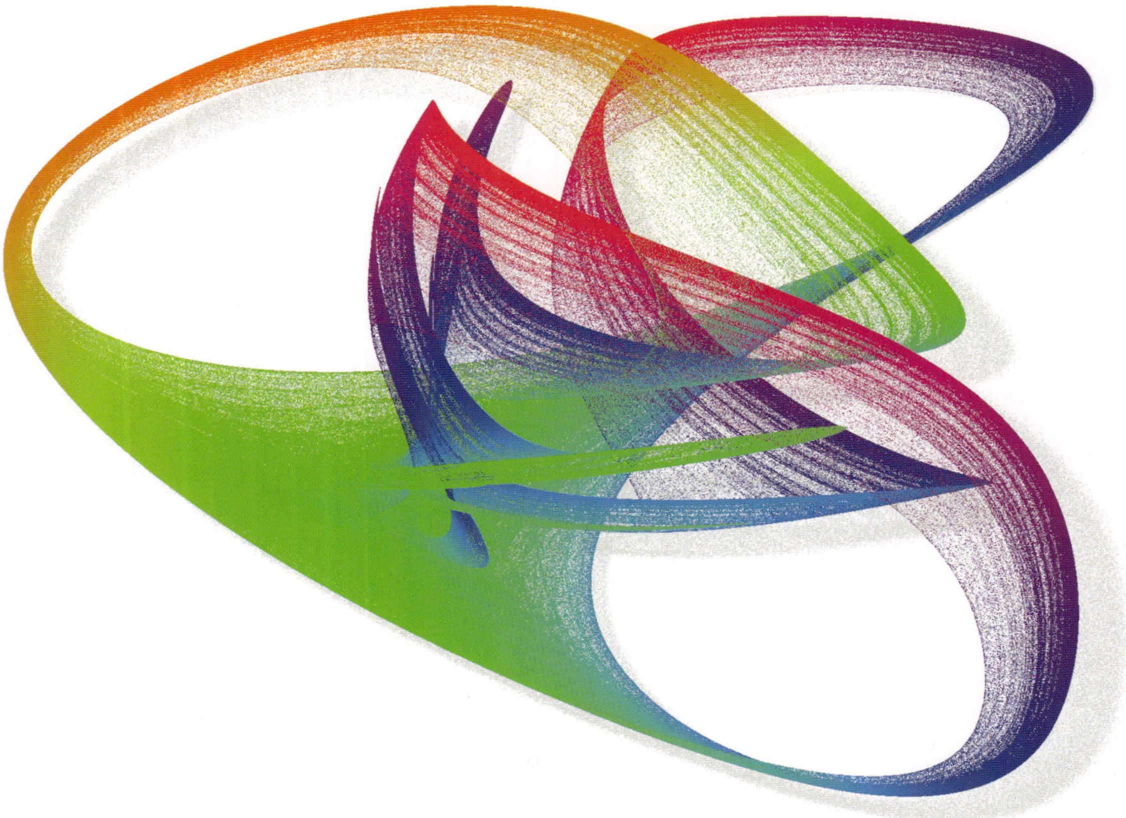

EVERY LAKE A LOON, AN EAGLE

Unearthly, we say, of the loon's call
        echoing through the dark,
its deep dive in black water, its low float,
        red eye, ringed neck,
diamonds of light like the lake in sun
        glittering along its back.

And when the pair came to our campsite,
        minor-keyed notes ascending,
falling, a music that threaded light
        through lake and mist to cloud,
we turn as though summoned:
        The male rose up before us
and beat his wings about his moon-white
        breast, a winged being
singing water and sky to the earth-bound,
        and the full moon rose.

I JOIN THE WOMEN OF CHURCHILL COLLEGE

Beside me, the barebreasted girl with braids,
a warrior's stance, gazes stonily across the hall;
her older sister, firm-bellied blue-marbled nude,
waits at the Fellows' back door, beside
the staffs of office. On the wall, a woman
sleepwalks through her charcoal portrait.
In the Senior Commons Room, another nude,
small, curls up in langorous bronzed sleep
by the stationery and blotting pad.
At dinner, the girl in the tartan skirt
bangs the dishes and snaps her oversize
rubber gloves, impatient to do the washing up
while I'm still eating my visitor's pudding.
The Fellows linger, talking The Critique
of Pure Reason, quantum tunneling.
Outside the Hall, opposite Churchill's bust
glowering in black depression, the last
of the beautiful nudes, life-size, sits and weeps,
her face in her hands.

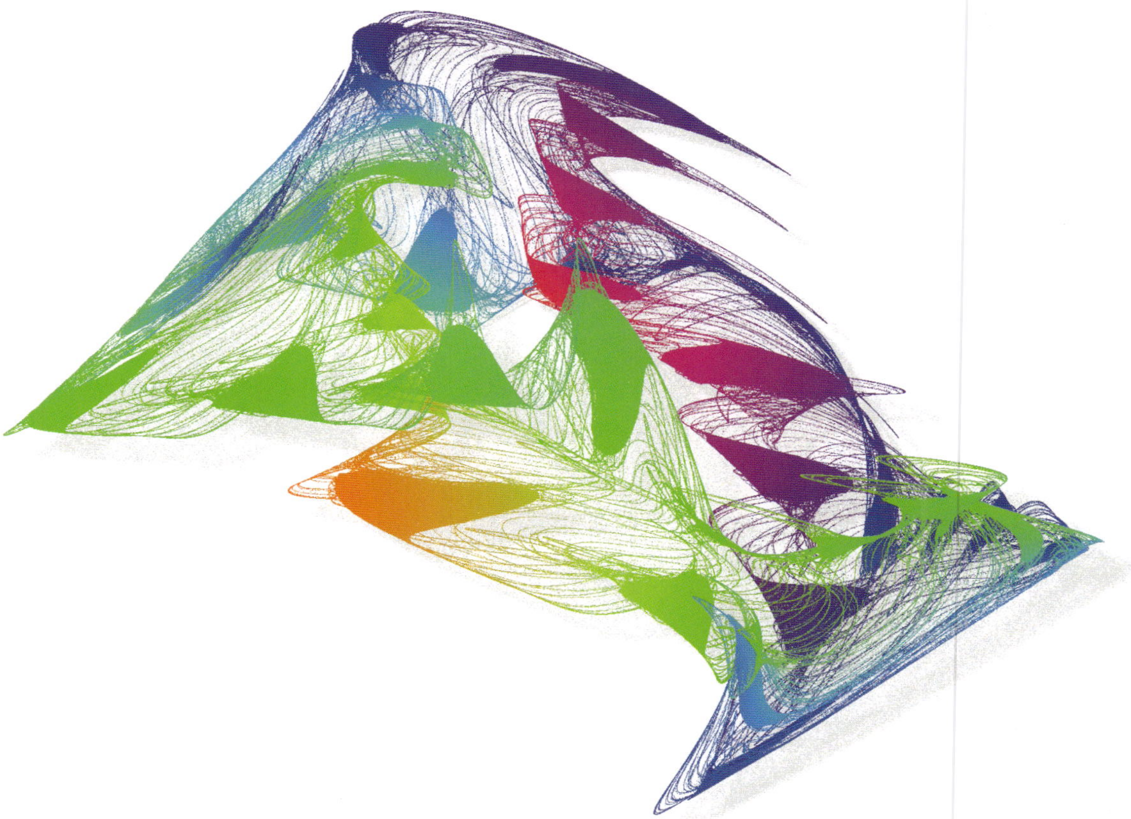

THE WOMAN WHO REMEMBERED NUMBERS

Every phone number from every house
she'd ever lived in; the phonebook
a nightmare to open, pages she couldn't forget;
house numbers, of course,
and license plates — the family's old Edsel,
the car that stopped in front of her
at the Midvale light last week;
a number — total revenue — glanced at
in the state budget book;
the shifting Dow Jones average
and temperature at the bank running
like rivers through her thoughts — and all
the constants: Avogadro's number;
the golden mean; pi, to as many digits
as she'd read. Books of logarithms
and measurement equivalents. Walking
by Resurrection Cemetery, so many dates
of birth and death. So much exact
and useless knowledge, colored and buzzy,
a feel in the body, varied scents. Hotel rooms,
locker numbers and combinations, running times
for her three-mile jog, through woods now,
trying to make a leafy green wildness,
an unnumbered lake, a lush meadow
of forgetfulness rise and flower in her head.

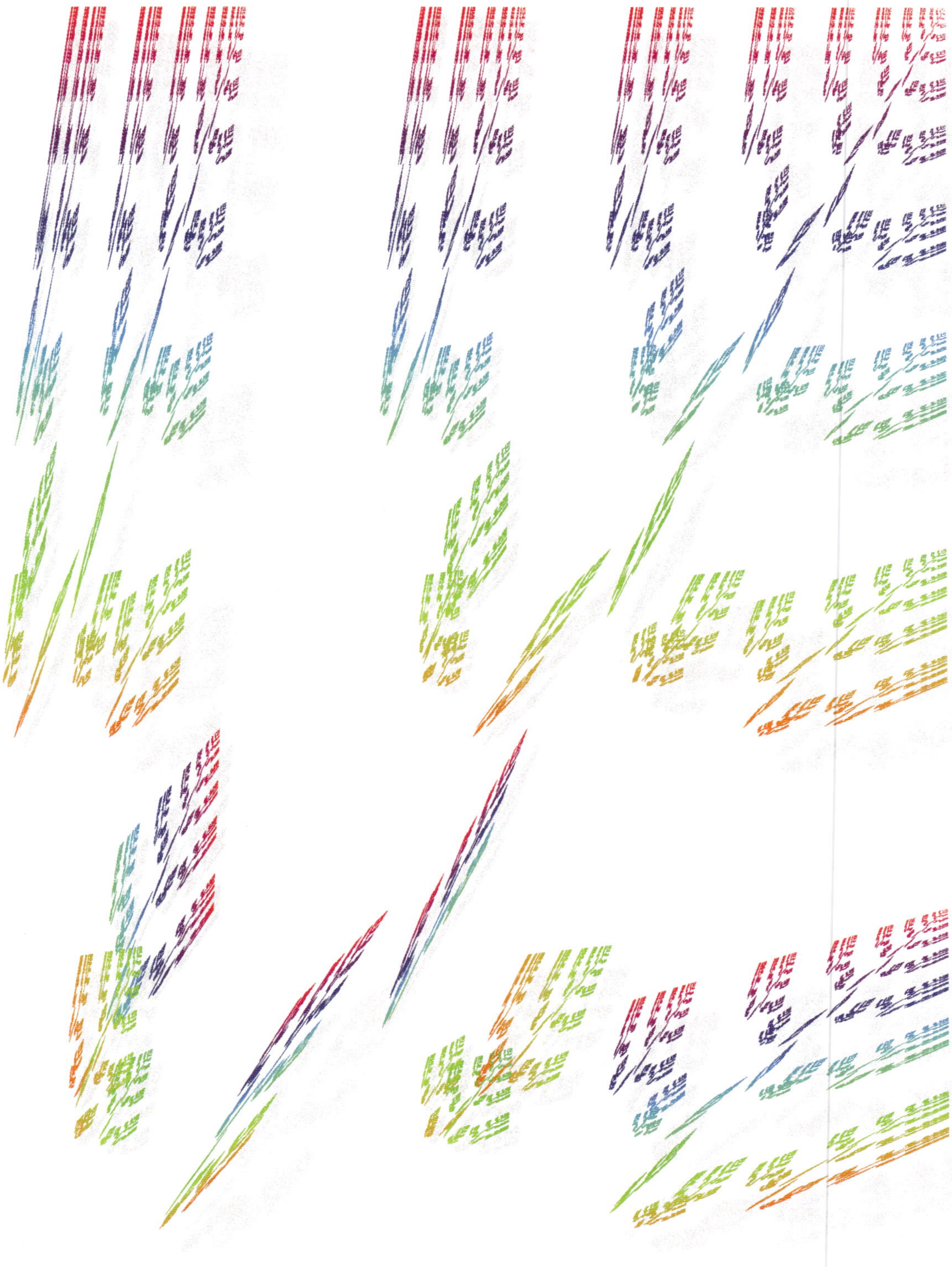

Flower Petals

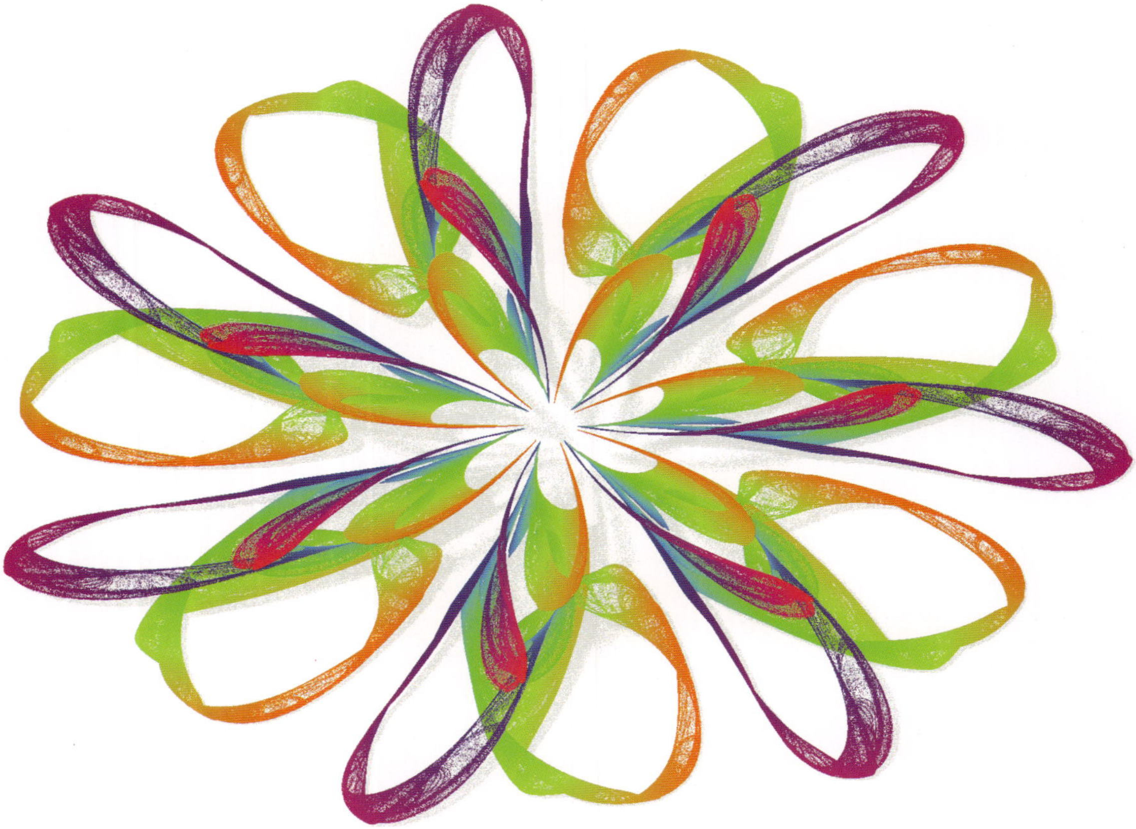

RANUNCULI

They are the orange of poppies, the pink
Of roses, the deep dusk of burgundy
Held up on corkscrewed green stems.

Tight layers of petals splay open,
Shading the light to apricot, peach,
Lemon along the tips, a slow curl to quill,

Reluctant fall — and the dark heart of metaphor
Ripens to raspberry, blackberry, holding
The essential, incomparable flower.

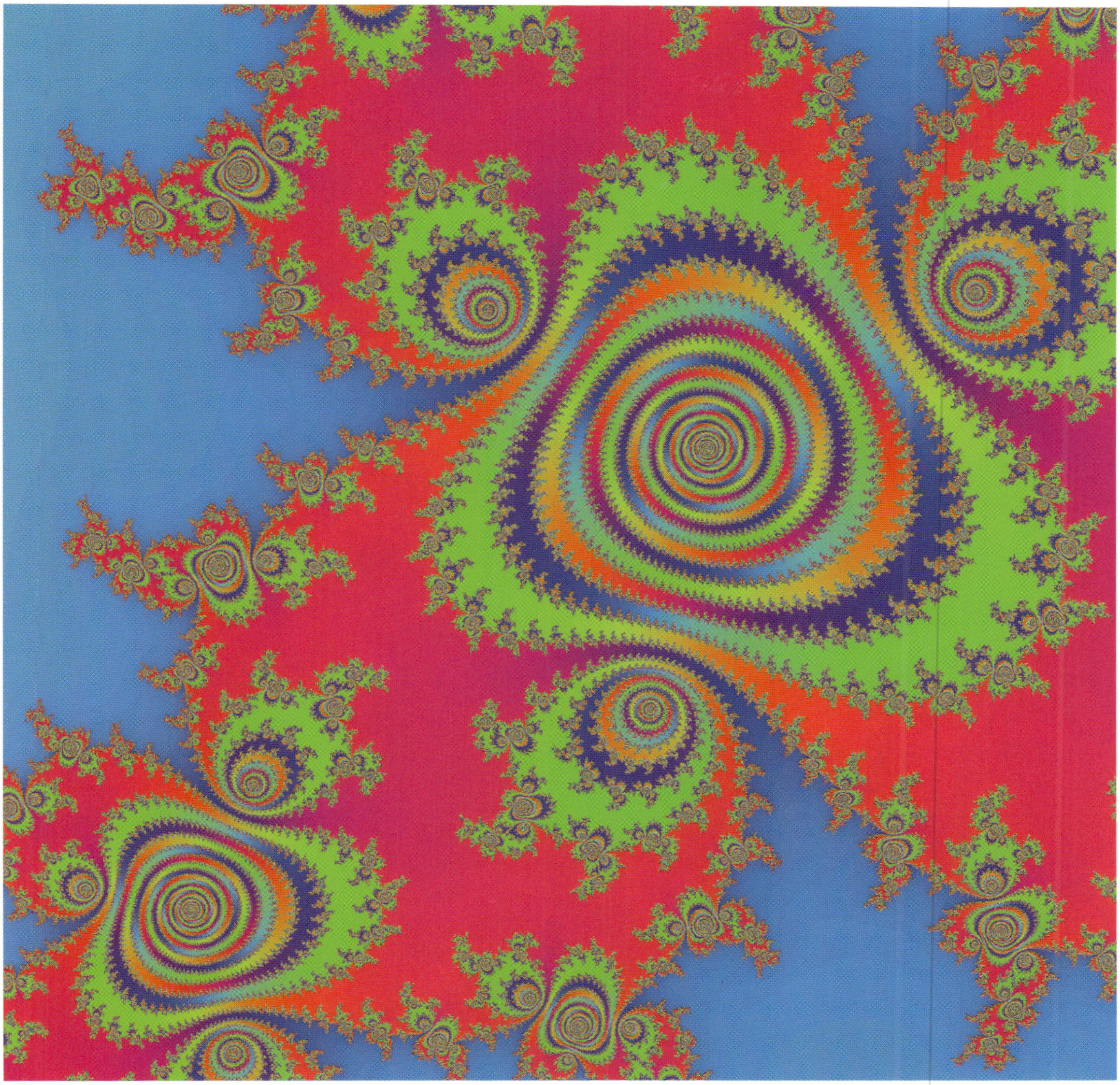

Symmetry

Symmetries play an important role in nature and in the mathematical theories that describe it. Heavenly bodies such as stars and planets have spherical symmetry, and humans and most other animals have bilateral symmetry. Flowers, starfish, diamonds, and snowflakes have symmetries with respect to rotation through specific angles. Fractals are symmetric with respect to changes in scale. Symmetries in the laws of physics with respect to displacements in space, time, and rotation lead to the conservation of linear momentum, energy, and angular momentum, respectively. The ubiquity of symmetries in nature probably accounts for their aesthetic appeal in computer art.

KAREN

Karen connects the butterflies in Mexico
to the students of English in Wisconsin and China,
brings trekking the Himalayas
and hitchhiking Africa to strolling the Cam,
the nuances of American accent
to foreign students, forges the link
of union politics and academic staff.

Karen collects animals, even those we didn't pet —
her office python that filed itself away
and emptied the building in the days
before bomb threats. And the ones we milked
and fed and petted — her goats, her cats, whose names
are legion: Chaos, Random, Minka, Orange Boy,
weave around our ankles.

Karen walks Ramey, her operatic companion dog,
with Jake, his redbone sidekick, sunrise in the parks,
nightly checking neighbors' yards, observes
monthly bring-your-dog-to-work day. Retiring
but not shy, she celebrates her sixtieth year, brings us
Hawaiian islands, rainforest tours, Danish house concerts,
sells us organic vegetables at the Regent Co-op.

Karen, our effervescent lover of folk dance
and nickelharpa, our maker of Folklore Village
community, our spiky redhead-blonde,
our tiger-glassed toe-ringed nose-jeweled friend,
who fearlessly speaks her mind, who raises a riot
of blossoms in every inch of yard — Karen's degree
of separation from the world must average 0.0.

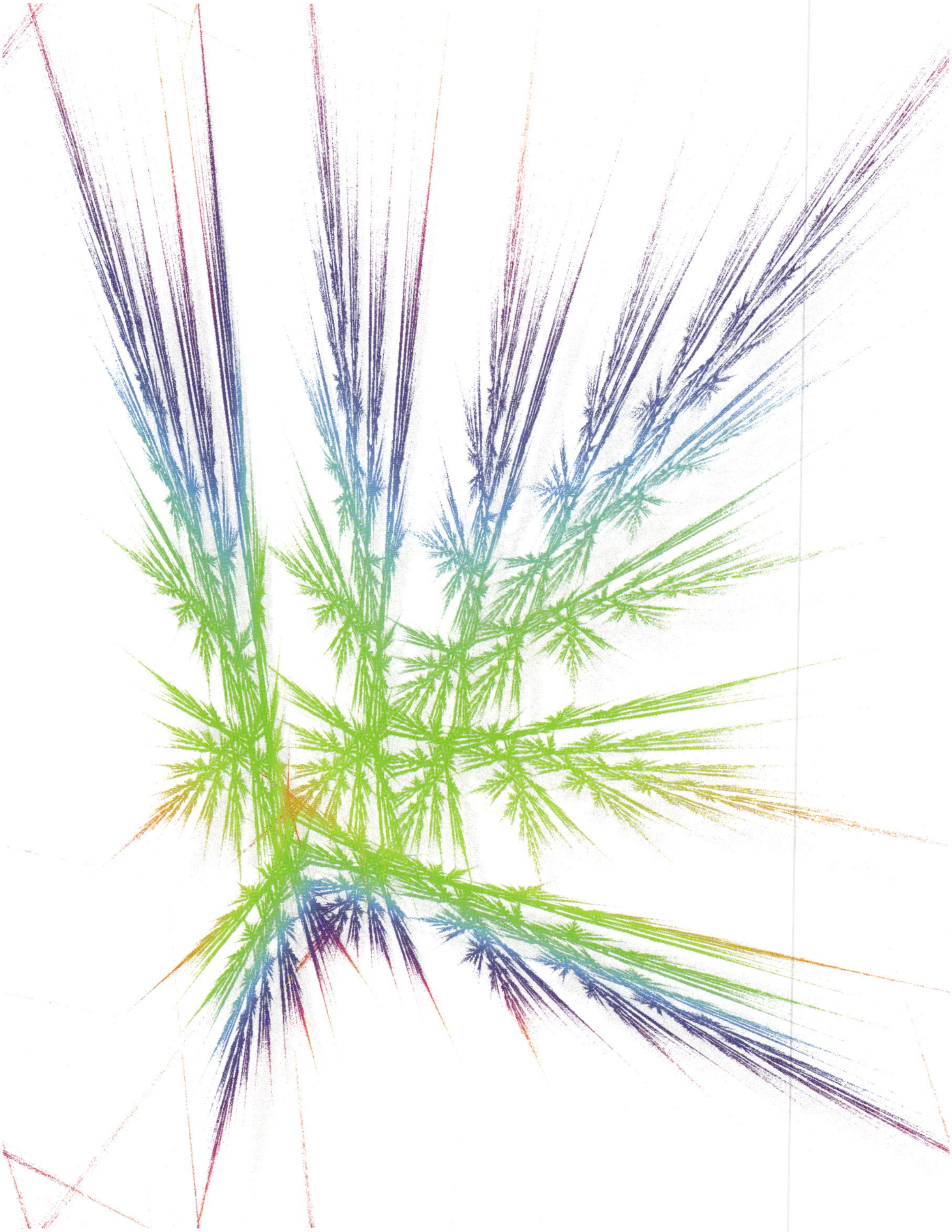

FINDING A WAY

Maybe you love birds, the way the cardinal
shines in sun on the topmost branch,
the quick intelligence of crows conversing
in the oaks on garbage day, the cranes'
trans-continental bugling as they follow
migratory routes through seventy-year lives,
and maybe you band together across
borders to watch out for them, though
some speak cardinal and some speak crow.

Maybe you love bowling, the heft of the ball,
the quick step up to the line, its spinning strike,
and find others of your kind to meet
weekly — clubs and leagues to cheer
each other on, shouts and jokes and fizzy drinks
with the bowling crowd, and maybe the kids
come too, and their friends, and your wife,
and in the loud reverberations of pins
and gutter balls you find you love your life.

Maybe you love dancing, the way
when you hold out a hand, she takes it in hers,
the way the whole line of you criss-cross steps,
snaking a grapevine pattern across the floor
to folk music from another place,
and you take lessons in square dance figures
or the resurrected Charleston and Lindy, learn
the turns of the Viennese waltz, the hiphop, the hambo,
the schottische, the hipswivel of mambo and salsa.

Maybe you love singing and find
a Friday night when all ages arrive at your house
with potluck squash and guitars, songbooks
and dulcimers, to sit on the floor and call out
City of New Orleans and *Julian of Norwich*
and *Waltzing with Bears* and *Roseville Fair*
and your voices rise and blend to rounds
and sea chanties, gospel and Dylan,
lullabies in every language.

Maybe you love the earth,
the way black Wisconsin soil crumbles
after the winter frost, engenders cool lettuce
and snow peas, hot peppers and tomatoes,
the sweet roots of beet and potato and rutabaga,
the plentiful squashes; your labor of digging, sowing,
weeding, and watering, the harvests, the feasts
that come after at the bountiful table you share,
the birds come to the hanging sunflower heads.

Whatever you love, begin there.

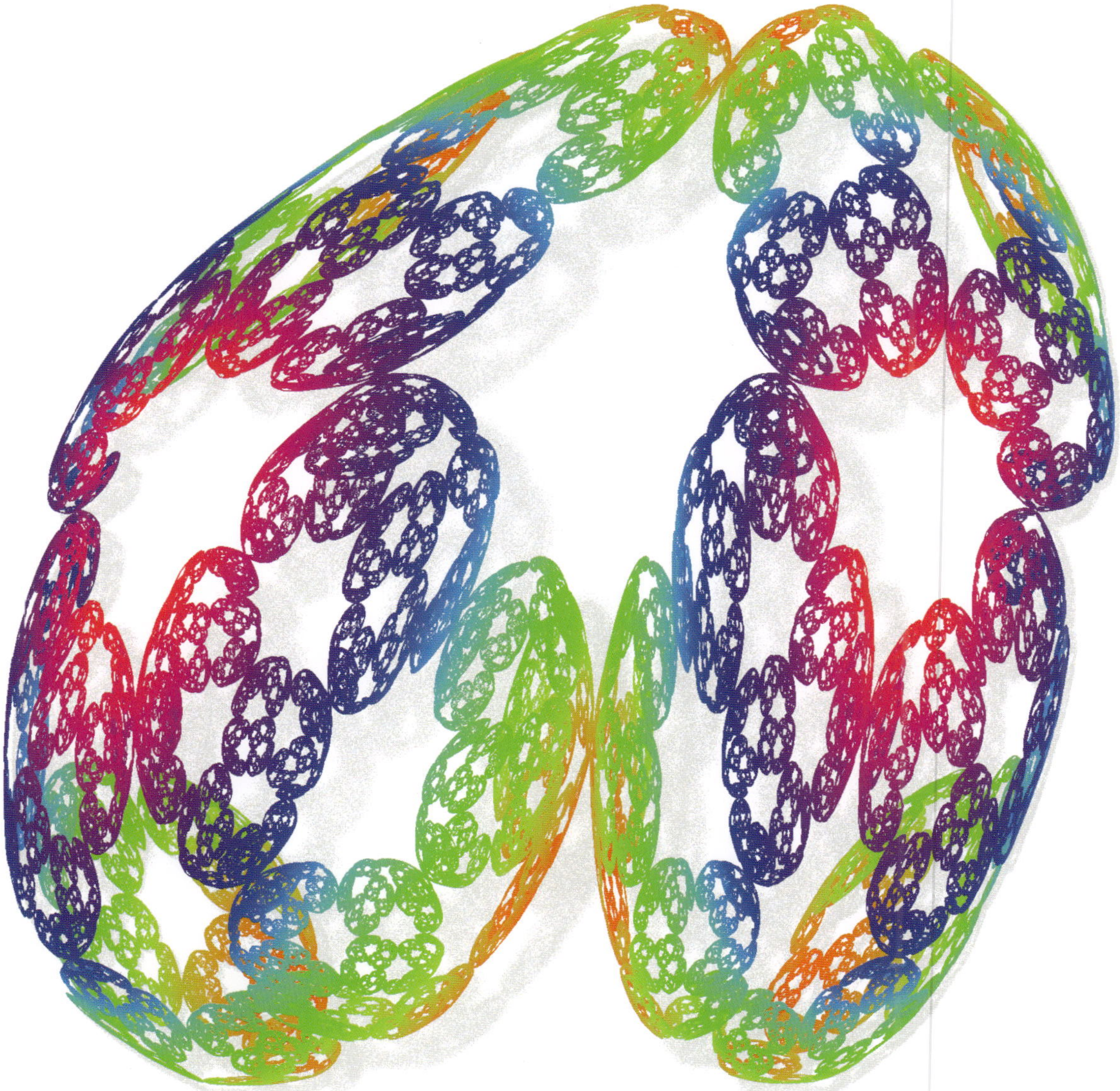

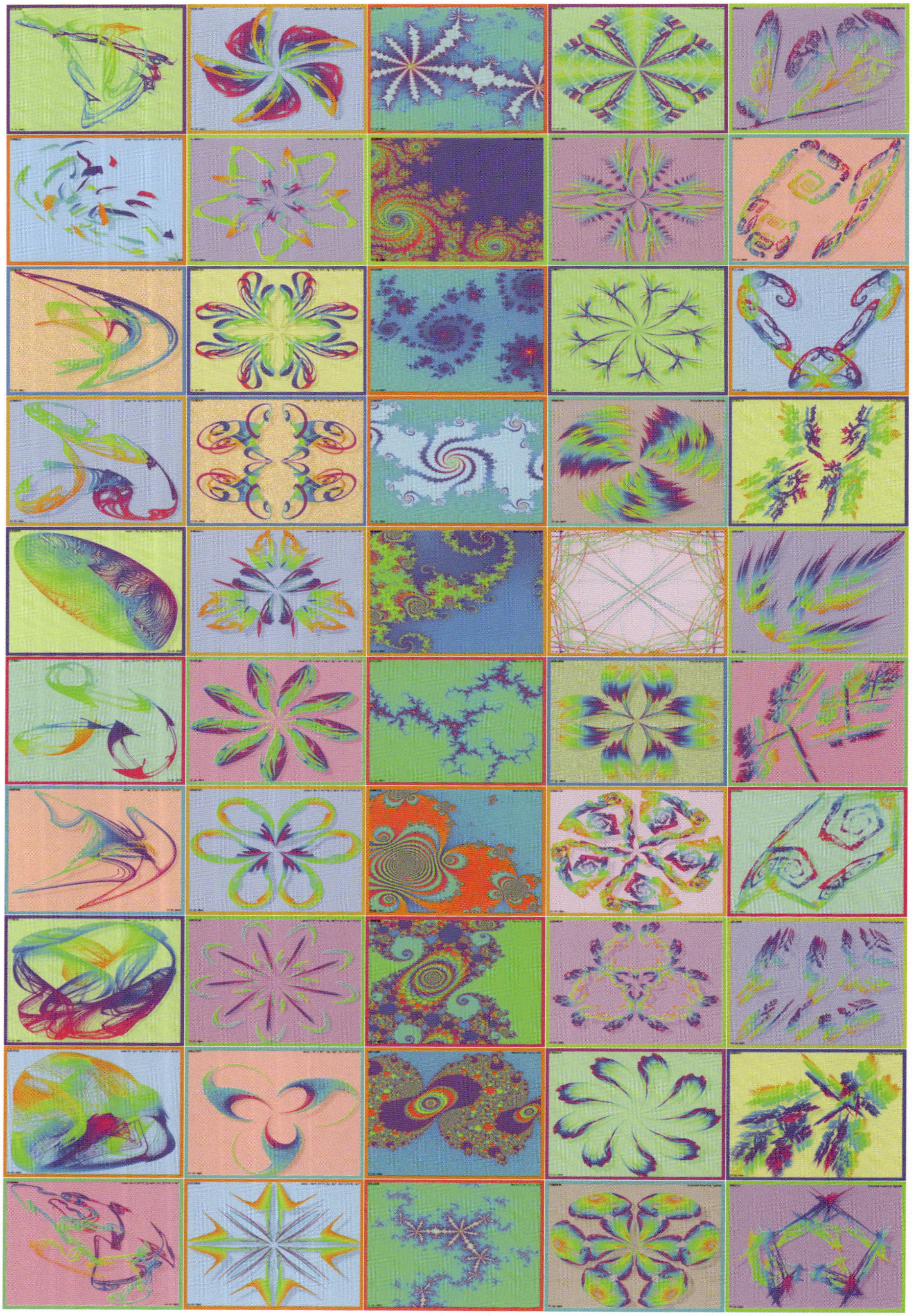

Appendix for the Mathematically Inclined

To the outsider, mathematics is a strange, abstract world of horrendous technicality, full of symbols and complicated procedures, an impenetrable language and a black art. To the scientist, mathematics is the guarantor of precision and objectivity. It is also, astonishingly, the language of nature itself. No one who is closed off from mathematics can ever grasp the full significance of the natural order that is woven so deeply into the fabric of physical reality. — Paul Davies (Australian astrobiologist)

The computer art in this book is of five types. This appendix will give the mathematics needed to understand how they were produced and how to replicate them.

1. Strange Attractors

The strange attractors are the result of several million iterations of the four-dimensional iterated map

$$X_{n+1} = a_1 X_n + a_2 X_n^2 + a_3 Y_n + a_4 Y_n^2 + a_5 Z_n + a_6 Z_n^2 + a_7 C_n + a_8 C_n^2$$

$$Y_{n+1} = X_n$$

$$Z_{n+1} = Y_n$$

$$C_{n+1} = Z_n$$

in which the successive iterates are plotted at a horizontal and vertical position given by X and Y, respectively. The height above the XY-plane is given by Z and is indicated by occlusion (larger Z values overwrite smaller ones) and by plotting a diffuse shadow below and to the right of each point displaced by a distance proportional to Z as if the attractor were illuminated by a distant source of light over your left shoulder. Points are plotted in a color C chosen from a rainbow palette of 248 colors with red, green, and blue intensity values (0 to 255) given by

$$Red = 127[1 + \cos(2\pi C)]$$

$$Green = 127[1 + \cos(2\pi C - 2\pi/3)]$$

$$Blue = 127[1 + \cos(2\pi C - 4\pi/3)]$$

where C has been scaled to the range of 0 to 1. For all cases, initial conditions are taken as $X_0 = Y_0 = Z_0 = C_0 = 0.05$, and the first few thousand iterates have been discarded to ensure that the orbit is on the attractor and to determine the range of the variables for plotting. For the images on the CD-ROM, a border color is chosen to be the same as the color used for the most pixels in the image, and the background color is taken as the half-intensity complement of the border color. The eight parameters a_1 through a_8 are coded into an eight-character string according to the ASCII value of each character A_n in the string using

$$a_n = 0.2(\text{asc}\, A_n - 77)$$

which allows values of $-2.4 \le a_n \le 2.4$ in steps of 0.2 for A_n = "A" to "Y" for a total of 25^8 or over a hundred billion combinations, a small fraction of which give visually interesting strange attractors [1–3].

2. Strange Attractor Icons

The strange attractors described above are too unstructured for some people's artistic taste. It is possible to constrain the equations so that they have certain symmetries [4, 5], but a simpler method is to introduce symmetry after the fact by replicating the image several times in a distorted wedge shape and assembling the wedges like the petals of a flower [6]. This is done in two steps, first converting X to a radius r, and Y to an angle θ using

$$r = (X - X_{\min})/2$$

$$\theta = 2\pi(Y_{\max} - Y)/n[(Y_{\max} - Y_{\min}) + s]$$

where s in a randomly chosen integer in the range 1 to n, n is the number of sectors (typically 2 to 9), $X_{\min}$ is the minimum value of X, $X_{\max}$ is the maximum value of X, and similarly for Y. Then the (r, θ) position is converted back to (X, Y) coordinates using

$$X = X_a + r\sin\theta$$

$$Y = Y_a + r\cos\theta$$

where X_a and Y_a are the X and Y coordinates corresponding to the center of the image, respectively. For even values of n, alternate sectors (those with s odd) are inverted by replacing θ with $2\pi/n - \theta$.

3. Iterated Function Systems

Iterated function systems [7] are a set of linear affine mappings here taken in a simplified form as

Map 1:

$$X_{n+1} = a_1 + a_2X_n + a_3Y_n + a_4Z_n$$

$$Y_{n+1} = X_n$$

$$Z_{n+1} = Y_n$$

$$C_{n+1} = Z_n$$

Map 2:

$$X_{n+1} = a_5 + a_6 X_n + a_7 Y_n + a_8 Z_n$$

$$Y_{n+1} = X_n$$

$$Z_{n+1} = Y_n$$

$$C_{n+1} = Z_n$$

The maps are chosen randomly at each iteration. Thus there are two attracting fixed points, one for each map, that compete to capture the orbit. The resulting orbit is random rather than chaotic, but the attractor is a deterministic fractal with self-similar structure on all scales. The simplified equations are chosen so that there are only eight parameters, a_1 through a_8, which are coded into an ASCII string using the same method as with the strange attractors. The initial conditions and colors are also chosen in the same way as for the strange attractors [8].

4. IFS Icons

The iterated function system images are made into symmetric icons in exactly the same way as the strange attractors are made into icons as described above.

5. Generalized Julia Sets

The usual Julia set is the boundary of the basin of attraction for the fixed points of the complex quadratic map $Z_{n+1} + Z_n^2 + c$, where $Z = X + iY$ is a complex variable and $c = a + ib$ is a complex parameter whose value determines the shape of the set [9]. The generalized Julia set [10] images in this book are from the two-dimensional cubic map

$$X_{n+1} = a_1 + a_2 X_n + a_3 X_n^2 + a_4 X_n^3 + a_5 X_n^2 Y_n + a_6 X_n Y_n - 3a_4 X_n Y_n^2 + a_7 Y_n - a_3 Y_n^2 - a_5 Y_n^3/3$$

$$Y_{n+1} = a_8 - a_7 X_n + a_6 X_n^2/2 - a_5 X_n^3/3 + 3a_4 X_n^2 Y_n + 2a_3 X_n Y_n + a_5 X_n Y_n^2 + a_2 Y_n + a_6 Y_n^2/2 - a_4 Y_n^3$$

where the peculiar coefficients result from satisfying the Cauchy-Riemann conditions

$$\partial F/\partial X = \partial G/\partial Y$$

$$\partial F/\partial Y = -\partial G/\partial X$$

with F and G given by

$$X_{n+1} = F(X_n, Y_n)$$

$$Y_{n+1} = G(X_n, Y_n)$$

This condition ensures that the boundary is a fractal, and it serendipitously gives eight independent coefficients that are coded into a string in the same way as for the strange attractors and iterated function systems above except using

$$a_n = 0.1(\text{asc}\, A_n - 77)$$

The colors are chosen by cycling through the same palette as above using the number of iterates required for the orbit to reach $X^2 + Y^2 > 10^6$, after which it is assumed to escape to infinity. Orbits that do not escape after 32000 iterations are assumed to be within the set and are colored white (or the background color for images on the CD-ROM). Initial conditions are taken as $X_0 = Y_0 = 0$, and the boundaries are set at -1 to 1 for both X and Y.

References

1. J. C. Sprott, How Common is Chaos?, *Physics Letters A* **173**, 21–24 (1993).
2. J. C. Sprott, Automatic Generation of Strange Attractors, *Computers & Graphics* **17**, 325–332 (1993).
3. J. C. Sprott, *Strange Attractors: Creating Patterns in Chaos*, M&T Books: New York (1993).
4. M. Field and M. Golubitsky, *Symmetry in Chaos: A Search for Pattern in Mathematics, Art, and Nature*, Oxford University Press (1992).
5. I. Stewart and M. Golubitsky, *Fearful Symmetry: Is God a Geometer?*, Blackwell: Oxford (1992).
6. J. C. Sprott, Strange Attractor Symmetric Icons, *Computers & Graphics* **20**, 325–332 (1996).
7. M. F. Barnsley, *Fractals Everywhere*, Academic Press: Boston (1988).
8. J. C. Sprott, Automatic Generation of Iterated Function Systems, *Computers & Graphics* **18**, 417–425 (1994).
9. H.-O. Peitgen and P. H. Richter, *The Beauty of Fractals: Images of Complex Dynamical Systems*, Springer: Berlin (1986).
10. J. C. Sprott and C. A. Pickover, Automatic Generation of General Quadratic Map Basins, *Computers & Graphics* **19**, 309–313 (1995).

Index of Images

Page	Image	Type
113	GHMNPSYD	JULIA
115	CLGIEPPB	JULIA
118	KMNTMPLF	IFS
119	MNMULPOV6	IFSICON
120	GDKEONQW	SA
121	JKNTLPNG	IFS
123	GSDXFXIX9	SAICON
124	KQPJMNMM	IFS
125	GPEYJTIV	SA
126	OLOFOLLK	IFS
127	OPLDNLNQ	IFS
128	MMQLLNMO	IFS
129	NPLTNMLT	IFS
130	CLMNLDNE	JULIA
131	MEGGJBLN8	SAICON
132	UFINUGNE8	SAICON
133	HAQENEJD	SA
134	KKLKLOON3	IFSICON
135	KIRCRKTE	SA
137	HFHELFPT	SA
138	JKMJLLNT	IFS
139	ELCCFLKD7	SAICON
141	XDNLGYMI	SA
142	ILKNMJMY	IFS
143	HLECJEID	SA
145	DKXPSTOH	JULIA
147	MMKTMLNV8	IFSICON
148	IRIXPKNN	SA
149	LKPNLIGJ	SA
151	MPMRLNNB	IFS
152	NEIUKVOL7	SAICON
153	OJKOLHWR	JULIA
155	MQNSLLMI	IFS
157	NKNWNOKA	IFS
158	HEIDLXJW	SA
158	HXNWQLKT	SA
158	GAPQKDOO	SA
158	HJGBMLNJ	SA
158	KNTHMPRF	SA
158	IHFBMKLG	SA

Page	Image	Type
158	JAKBPRNW	SA
158	LLEHKJML	SA
158	PESHNNRA	SA
158	DSIRLYNR5	SAICON
158	LLJOHPLL5	SAICON
158	MHOKRTJE8	SAICON
158	IOJWKUIX4	SAICON
158	TEVFPIQE9	SAICON
158	VARAOFRC6	SAICON
158	XDOIVECW9	SAICON
158	KAKJJIPR3	SAICON
158	KSSIINRJ8	SAICON
158	DHTPHMKO	JULIA
158	IYUOWUOU	JULIA
158	KHUJWBRK	JULIA
158	LEMUPQGM	JULIA
158	XJJNUKEV	JULIA
158	HSDJODHO	JULIA
158	GHMNPSYD	JULIA
158	HBHGOHMN	JULIA
158	HETTLQNP	JULIA
158	MAPJORMA	JULIA
158	MQMDLLMN8	IFSICON
158	KQOOJMNM	IFSICON
158	JOMPSLLL9	IFSICON
158	KPODLPNX3	IFSICON
158	FKMIJOMK8	IFSICON
158	NNJMJNMB8	IFSICON
158	KMKDMJLK5	IFSICON
158	KOOWQNLA9	IFSICON
158	LKOBNOLN6	IFSICON
158	OPKVNLMS	IFS
158	PJOIONLT	IFS
158	PLLTJONE	IFS
158	OLOFOLLK	IFS
158	NPMTJOON	IFS
158	RIPWPKMV	IFS
158	RHOYONLK	IFS
158	NPLLMNNH	IFS
158	OKMKRIPJ	IFS

Test Your Understanding

Scientists often borrow words from ordinary language and give them a special technical meaning that does not always coincide with their common use. Test your understanding of some of the terms used throughout this book by choosing the definition closest to the technical meaning.

1. **Dynamical system (*page 2*)**
    a. A system whose variables change in time
    b. A system whose parameters change in time
    c. A system with a lot of energy
    d. A system with great influence on other systems

2. **Complex system (*page 5*)**
    a. A system governed by complicated rules
    b. A system that is highly nonlinear
    c. A system with many variables
    d. A system that exhibits chaos

3. **Linear system (*page 12*)**
    a. A system in which every effect has a single cause
    b. A system whose behavior follows a straight line
    c. A system with a single dimension or variable
    d. A system in which the output is proportional to the input

4. **Set of measure zero (*page 14*)**
    a. All the things that can happen with vanishing probability
    b. All of the solutions of a system of equations
    c. All occurrences of zero in the solutions of a set of equations
    d. All of the equations whose solutions are zero

5. **Nonlinear system (*page 17*)**
    a. A system in which effects have multiple causes
    b. A system whose whole is not equal to the sum of its parts
    c. A system that exhibits chaos
    d. A system with many variables

6. **Three-body problem (*page 21*)**
    a. Biological survival of the fittest
    b. The motion of three bodies with mutual attraction or repulsion
    c. The social interactions of three friends
    d. The riddle of a multiple homicide

7. **Control parameter (*page 23*)**
   a. A variable that is controlled by a dynamical system
   b. A variable that couples one system to another
   c. An object that stops a system from oscillating
   d. A knob that determines the behavior of a system

8. **Time series (*page 26*)**
   a. A row of digital clocks
   b. A sequence of measurements at equal time intervals
   c. A sequence of successive events
   d. A competition of fast moving variables

9. **Soliton (*page 30*)**
   a. A system that does not interact with adjacent systems
   b. A linear wave that propagates with distortion
   c. A nonlinear wave that propagates without distortion
   d. A wave that does not interact with objects that it encounters

10. **State space (*page 39*)**
   a. The space whose axes are the dynamical variables
   b. The space occupied by a particular state of the system
   c. The space in which one can explicitly state the behavior
   d. The space under control of a single entity

11. **Attractor (*page 50*)**
   a. An object that exerts an attracting force on another object
   b. One of the poles of a magnet
   c. The states which a dynamical system approaches after a long time
   d. A constraint that limits the possible behaviors

12. **Entropy (*page 52*)**
   a. A measure of the disorder in a system
   b. A measure of the number of degrees of freedom in a system
   c. A measure of the chaos in a system
   d. A measure of the complexity of a system

13. **Equilibrium (*page 54*)**
   a. A system that is immune to perturbations
   b. A system in which the forces are balanced
   c. A state to which the dynamics always returns
   d. The normal state of an object subject to its environment

14. **Strange attractor (*page 67*)**
   a. An object whose shape is unpredictable
   b. An object that attracts another dissimilar object
   c. An object that attracts another similar object
   d. A fractal produced by a chaotic process

15. **Bifurcation (*page 72*)**
   a. A qualitative change in the behavior of a system after a set amount of time
   b. A splitting into two branches
   c. A system that oscillates between two states
   d. A qualitative change in the behavior of a system when a parameter reaches a critical value

16. **Phase transition (*page 75*)**
   a. A change in the nature of a system when a parameter reaches a critical value
   b. A shift in the time at which an oscillating system reaches its peak
   c. A change in the shape of an object
   d. A change in the nature of a system after a certain time has lapsed

17. **Dimension (*page 83*)**
   a. The number of adjustable parameters a system has
   b. The degree of nonlinearity a system has
   c. The number of dynamical variables a system has
   d. The size of the real space in which the dynamics occur

18. **Routes to chaos (*page 86*)**
   a. The means by which chaos was discovered
   b. Methods for finding chaos in experimental data
   c. A sequence of bifurcations the last of which is chaotic
   d. The succession of points along a chaotic trajectory

19. **Period doubling (*page 87*)**
   a. A bifurcation in which the period suddenly becomes twice as great
   b. A system that switches between two coexisting attractors
   c. The mechanism responsible for chaos in strange attractors
   d. A bifurcation in which the size of the oscillation suddenly doubles

20. **Self-organization (*page 90*)**
   a. The periodic behavior of an apparently chaotic system
   b. The tendency for a random system to develop ordered structure
   c. Order imposed by the rules that govern a system's dynamics
   d. Then tendency for an ordered system to exhibit randomness

21. Self-organized criticality (*page 95*)

a. The tendency of complex systems to exhibit scale invariance
b. The tendency of a critical system to exhibit chaos
c. The tendency of a chaotic system to develop ordered structure
d. The tendency of an ordered system to behave randomly

22. Chaos (*page 98*)

a. Complete disorder
b. Sensitive dependence on parameters
c. Heterogeneous agglomeration
d. Sensitive dependence on initial conditions

23. Simulated annealing (*page 99*)

a. A method for smoothing a mathematical function
b. A way to make a metal softer
c. A way to make a metal stronger
d. A computer method for optimizing a quantity

24. Determinism (*page 100*)

a. The future is uniquely determined by the past
b. The process is completely predictable
c. Small changes in initial conditions have little effect
d. The behavior of the system cannot be controlled

25. Randomness (*page 101*)

a. The behavior of a chaotic system
b. The result of a deterministic dynamical process
c. The behavior of a complex system
d. The result of a process governed by pure chance

26. Sonification (*page 105*)

a. A method for studying dynamics by listening to the sound produced
b. Imposing a hierarchy on a system of equations
c. The analysis of complex sounds
d. A method for producing fractal music

27. Lyapunov exponent (*page 106*)

a. The growth rate of a replicating system
b. A measure of the space-filling nature of a fractal
c. A measure of sensitive dependence on initial conditions
d. A measure of the complexity of a system

28. Iterated function system (*page 110*)

a. A set of functions that are repeatedly iterated
b. A function whose iterative solutions are chaotic
c. A system whose function is to facilitate iteration
d. The attractor for a set of rules applied repeatedly in every possible order

29. Cellular automaton (*page 111*)

a. A small robot
b. A computer model with cells whose contents evolve by simple rules
c. An automated communication device
d. A system that spontaneously switches between states

30. Julia set (*page 113*)

a. The basin boundary for bounded solutions in parameter space
b. A collection of attractors for the Julia map
c. A map of all the Mandelbrot sets
d. The basin boundary for bounded solutions in the space of initial conditions

31. Basin of attraction (*page 114*)

a. The set of initial conditions that approach the attractor
b. A collection of attractors
c. The region of parameter space whose solutions approach the attractor
d. The set of initial conditions that produce unbounded solutions

32. Mandelbrot set (*page 117*)

a. A map of all the Julia sets
b. The basin of attraction for unbounded solutions
c. A chaotic system discovered by Benoit Mandelbrot
d. A random fractal used to produce computer art

33. Hysteresis (*page 120*)

a. A system that does not return to its initial state when time is reversed
b. The failure to reproduce a behavior when a parameter change is reversed
c. The excitation of an atom from its ground state to an excited state
d. Unexpected and surprising behavior of a dynamical system

34. Iteration (*page 123*)

a. Repeated application of a mathematical procedure
b. A method of counting
c. Successive approximations of a solution
d. A computer procedure that acts on itself

35. **Butterfly effect (*page 131*)**
    a. Turbulence produced by fluctuating organisms
    b. Irregular oscillations of a dynamical system
    c. Behavior of a complex system
    d. Sensitive dependence on initial conditions

36. **Fractal (*page 140*)**
    a. An object that is broken into many pieces
    b. An object that is produced by a chaotic process
    c. An object that contains miniature copies of itself
    d. An object whose boundary is ragged

37. **Cantor set (*page 142*)**
    a. A finite collection of points with infinite measure
    b. An infinite collection of points with zero measure
    c. An infinite collection of points with finite measure
    d. A finite collection of points with finite measure

38. **Zipf's law (*page 144*)**
    a. The principle of competitive exclusion
    b. The law that energy is conserved
    c. The observation of scale invariance in language
    d. Nature abhors a vacuum

39. **Symmetry (*page 153*)**
    a. Invariance with respect to some mathematical transformation
    b. Looking the same front and back
    c. Looking the same right and left
    d. Two systems that behave identically

Answers:

1. a	11. c	21. a	31. a
2. c	12. a	22. d	32. a
3. d	13. b	23. d	33. b
4. a	14. d	24. a	34. a
5. b	15. d	25. d	35. d
6. b	16. a	26. a	36. c
7. d	17. c	27. c	37. b
8. b	18. c	28. d	38. c
9. c	19. a	29. b	39. a
10. a	20. b	30. d	

For Further Reading

- Briggs, J. *Fractals: The Patterns of Chaos* (New York: Touchstone, 1992). A non-mathematical but detailed explanation of fractals and chaos with many high-quality drawings and photographs.
- Field, M. and M. Golubitsky. *Symmetry in Chaos: A Search for Pattern in Mathematics, Art, and Nature* (Oxford: Oxford University Press, 1992). A spectacularly illustrated book that includes computer-generated images of flowers, graphic logos, motifs, and other artistic mathematical patterns.
- Gleick, J. *Chaos: Making a New Science* (New York: Viking Penguin, 1987). This best-selling, historical, and non-technical account is the starting point for anyone who wants to understand why chaos has excited the imagination of the scientist and nonscientist alike.
- Mandelbrot, B. B. *The Fractal Geometry of Nature* (San Francisco: W. H. Freeman, 1982). An extended essay by the father of fractals, this was the seminal work that brought to the attention of the nonspecialist the ubiquity of fractals in nature.
- McGuire, M. *An Eye for Fractals* (Reading, MA: Addison-Wesley, 1991). This graphic and photographic essay by a physicist and amateur photographer contains beautiful black and white photographs of natural fractals in the style of Ansel Adams, with a simple mathematical description of fractals.
- Peitgen, H. O. and P. H. Richter. *The Beauty of Fractals: Images of Complex Dynamical Systems* (New York: Springer-Verlag, 1986). This beautiful, colorful exhibit of computer art emphasizes the Mandelbrot and Julia sets.
- Pickover, C. A. *The Loom of God: Mathematical Tapestries at the Edge of Time* (New York: Plenum, 1997). This is one of a number of fascinating books by Clifford A. Pickover dealing with fractals, mathematics, computers, science, history, and other related topics.
- Porter, E. and J. Gleick. *Nature's Chaos* (New York: Viking Penguin, 1990). This book of art combines photographs of natural fractals by Porter with a simple, almost poetic, explanation of chaos and fractals by Gleick.
- Sprott, J. C. *Strange Attractors: Creating Patterns in Chaos* (New York: M&T Books, 1993). This book by one of the authors contains explanations and computer programs that allow the reader to search for visually interesting strange attractors and display them in various ways.
- Sprott, J. C. *Chaos and Time-Series Analysis* (Oxford: Oxford University Press, 2003). This slightly technical book by one of the authors covers all the material here and much more in considerably greater depth, requiring an elementary knowledge of calculus.
- Sprott, J. C. and G. Rowlands. *Chaos Demonstrations* (North Carolina State University, Raleigh, NC: Physics Academic Software, 27695-8202). An IBM PC program by one of the authors and a colleague that provides a simple way to learn about chaos, fractals, and related phenomena. It comes with an 84-page user's manual and 3-D glasses.
- Stewart, I. and M. Golubitsky. *Fearful Symmetry: Is God a Geometer?* (Oxford: Blackwell, 1992). This sequel to an earlier book by Stewart deals with symmetry in nature, art, and science and provides computer programs for producing symmetrical patterns of considerable beauty.

Contents of the CD-ROM

The CD-ROM that accompanies this book contains 1000 images in 1024 × 768 resolution plus 100 images in 4400 × 3200 resolution with colored backgrounds and borders, a sample of which is shown below. It also contains audio readings of the poems in Chapter 1 by the author, Robin Chapman, a slide show, an interactive quiz, and links to Web resources.

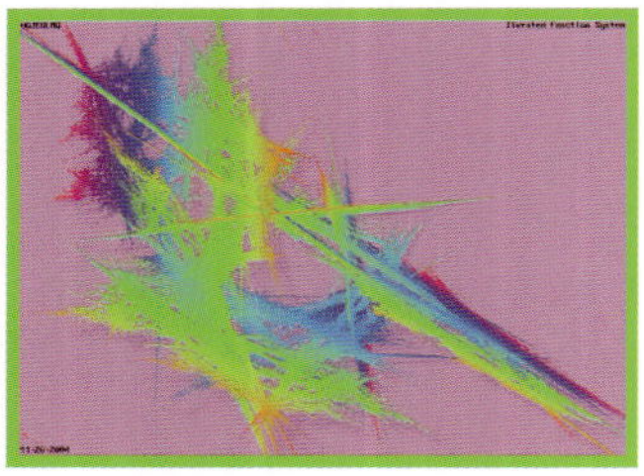
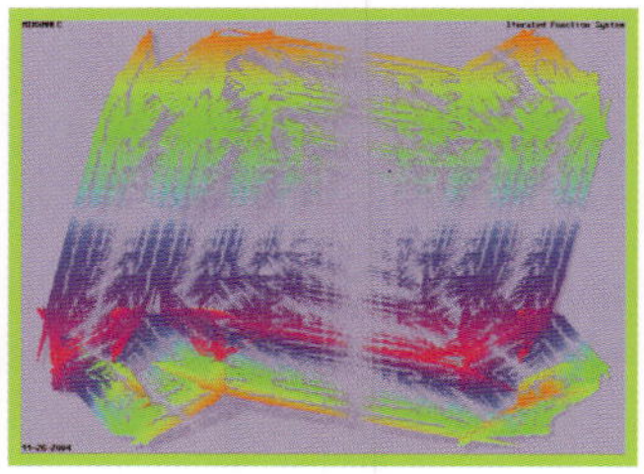
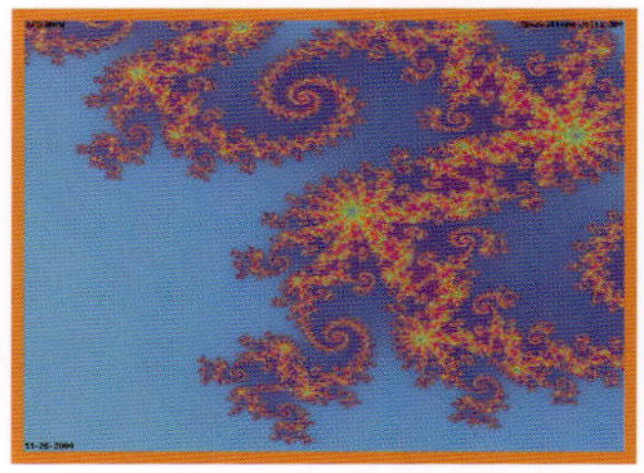
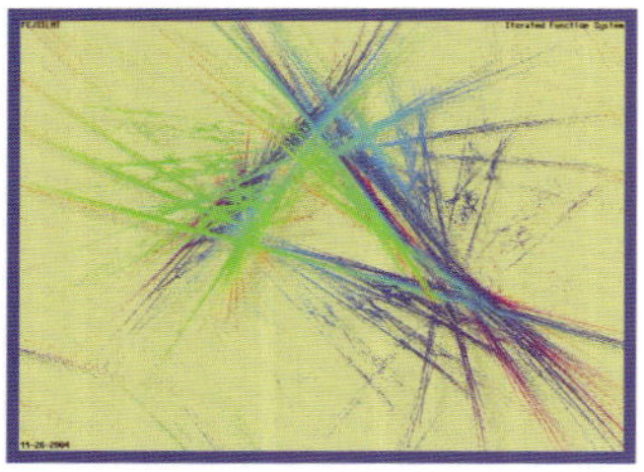
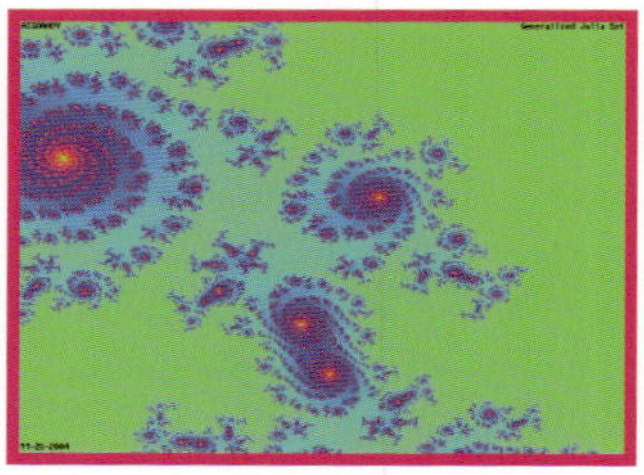
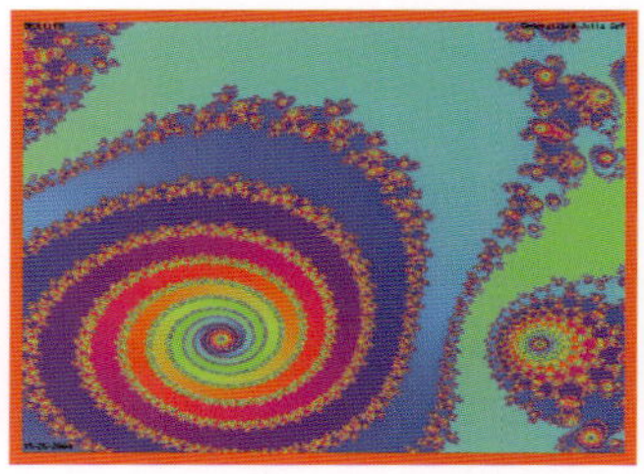
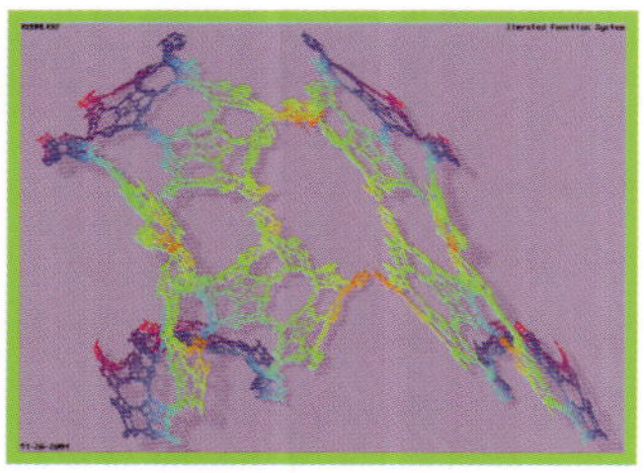
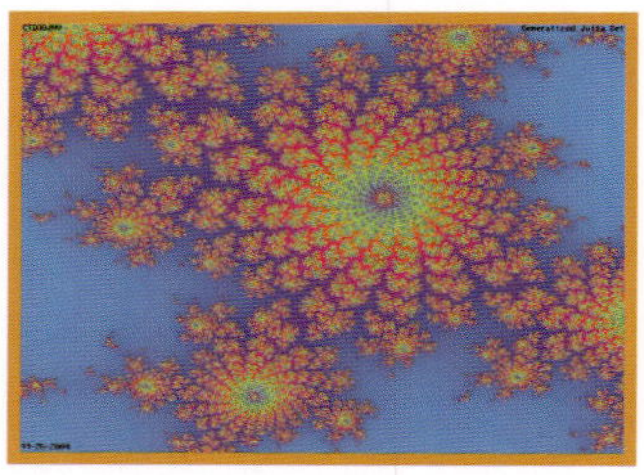
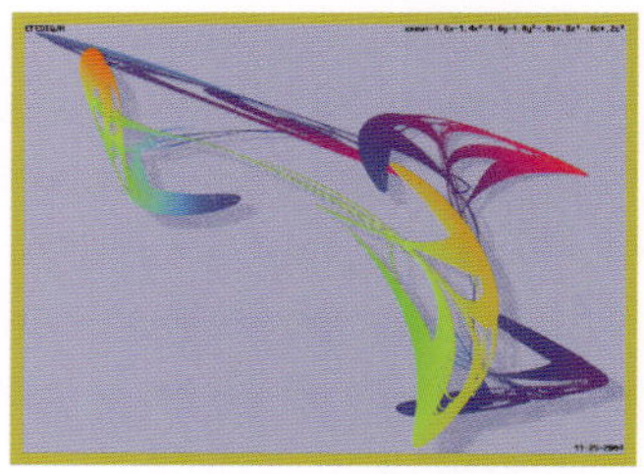
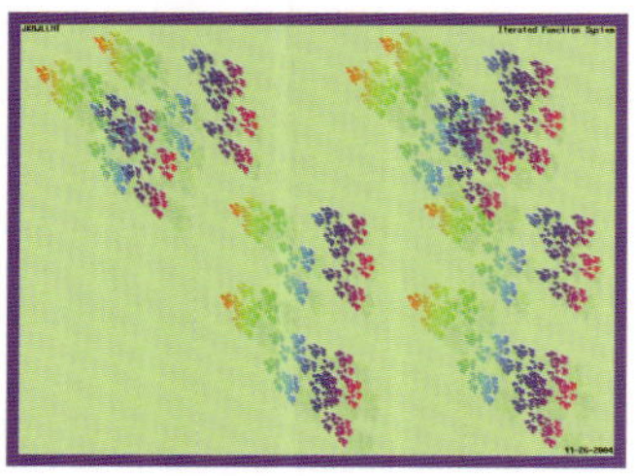
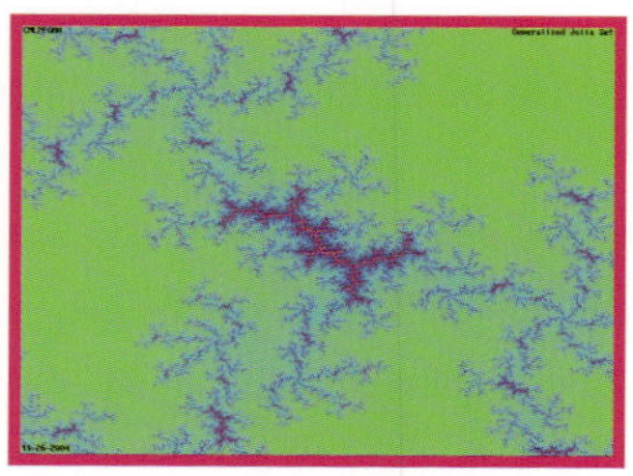
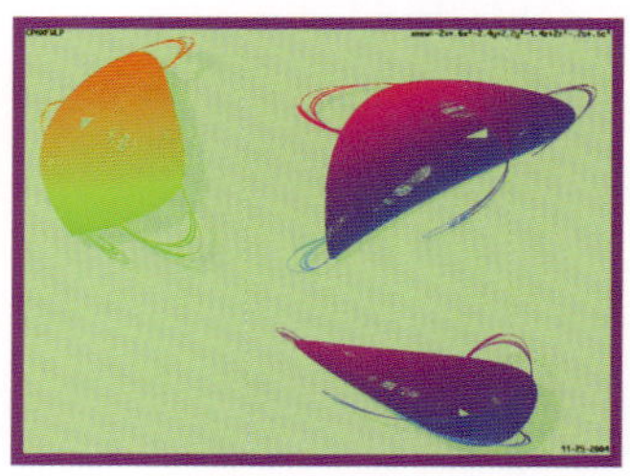
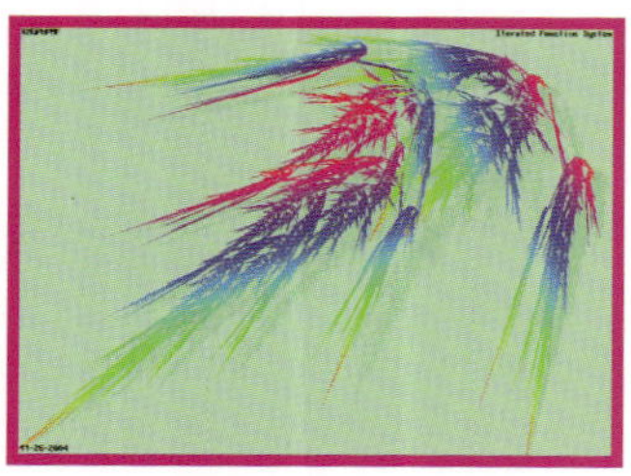
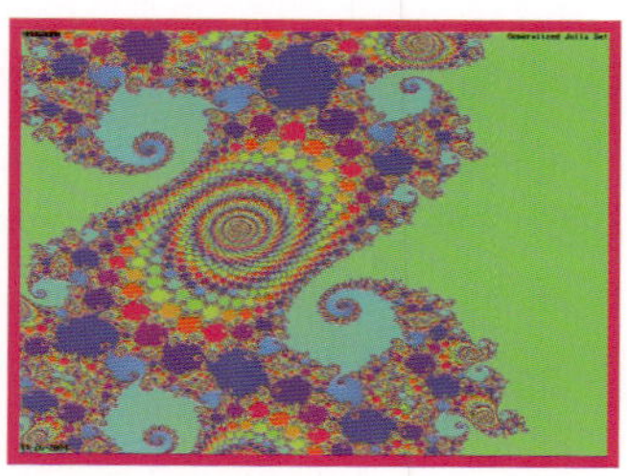
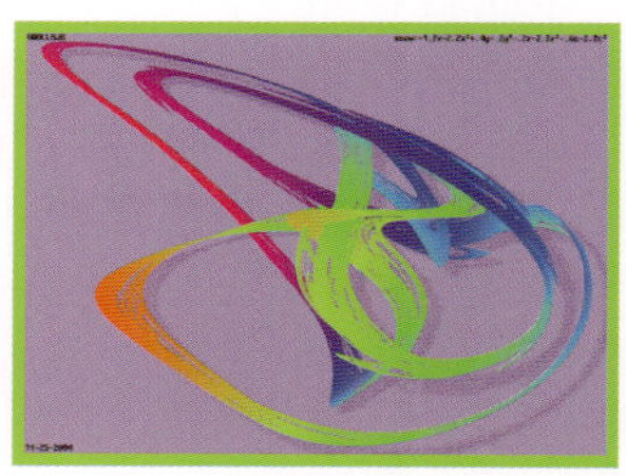

About the Authors

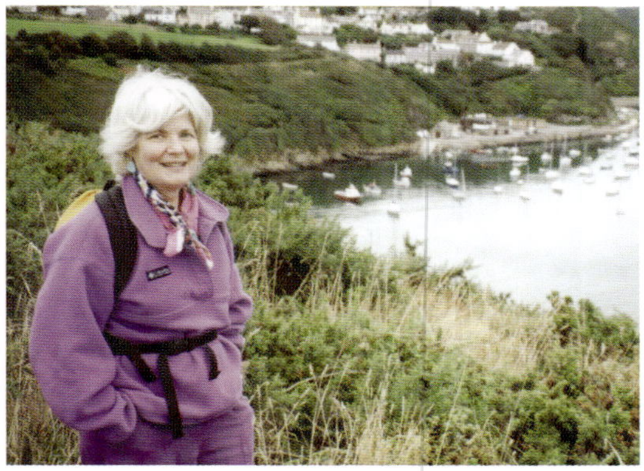

Robin Chapman was born in Washington, DC and raised in Oak Ridge, Tennessee. She earned a bachelor's degree in psychology from Swarthmore College and a Ph.D. from the University of California at Berkeley. She is currently Emerita Professor of Communicative Disorders at the University of Wisconsin and a principal investigator at the Waisman Center. Her current research, funded by the National Institute of Child Health and Human Development, with additional funding from the National Down Syndrome Society, focuses on language and cognitive development in children and adolescents with Down syndrome. She is co-author of SALT, a computer program to analyze language transcripts, and a Fellow of the American Speech, Language, and Hearing Association. Her poems have appeared in *Poetry*, *The American Scholar*, *The Southern Review*, and *The Hudson Review*, among other journals. A recipient of two Individual Artist Development grants from the Wisconsin Arts Board, she has published two books and five chapbooks of poems, including *The Way In*, which won the Posner Poetry Award from the Council for Wisconsin Writers, and *The Only Everglades in the World*. She lives with her husband, Will Zarwell, in Madison, Wisconsin.

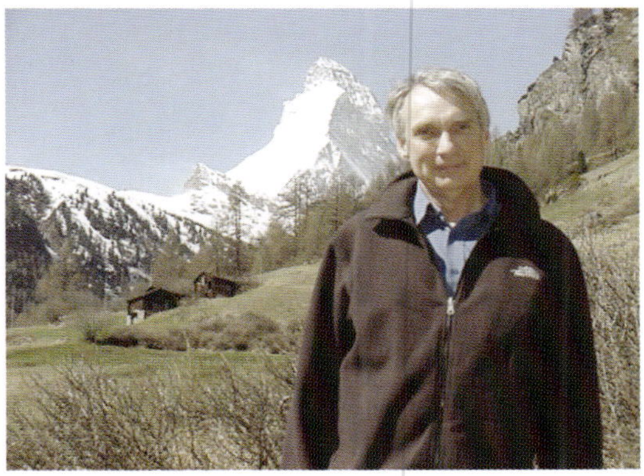

Julien Clinton Sprott was born and raised in Memphis, Tennessee. He earned a bachelor's degree in physics from MIT and a Ph.D. from the University of Wisconsin, after which he spent two years at the Oak Ridge National Laboratory in Tennessee, before returning to the University of Wisconsin as a Professor of Physics. He spent most of his career developing magnetic fusion energy. He has published about 300 papers on plasma physics, chaos, fractals, and complexity, and has written several books, including *Introduction to Modern Electronics*, *Numerical Recipes and Examples in BASIC*, *Strange Attractors: Creating Patterns in Chaos*, *Chaos and Time-Series Analysis*, and *Physics Demonstrations*. In 1984 he began a program called *The Wonders of Physics*, in which he has made about 200 public lecture/demonstrations to a total audience in excess of 50,000. He has produced twenty-two videos and written four commercial computer programs for demonstrating principles of physics, chaos, and complex systems. He has won several awards for his work in public science education, including the John Glover Award from Dickinson College, the Van Hise Outreach Teaching Award from the University of Wisconsin, and a lifetime achievement award from the Wisconsin Association of Physics Teachers. His award-winning Web site is at http://sprott.physics.wisc.edu/.

Index